Lab*Oratory*

Lab*Oratory*

Speaking of Science and Its Architecture

Sandra Kaji-O'Grady and Chris L. Smith

The MIT Press
Cambridge, Massachusetts
London, England

This book was set in Stone Serif and Stone Sans by Westchester Publishing Services. Printed and bound in the United States of America.

Library of Congress Cataloging-in-Publication Data

Names: Kaji-O'Grady, Sandra, author. | Smith, Chris (Chris L.), author.
Title: LabOratory : speaking of science and its architecture / Sandra Kaji-O'Grady and
    Chris L. Smith.
Description: Cambridge, MA : The MIT Press, 2019. | Includes bibliographical references
    and index.
Identifiers: LCCN 2019007762 | ISBN 9780262043328 (hardcover : alk. paper)
Subjects: LCSH: Laboratories. | Architecture and science.
Classification: LCC NA6751 .K35 2019 | DDC 720.1/05—dc23
LC record available at https://lccn.loc.gov/2019007762

10   9   8   7   6   5   4   3   2   1

Contents

Acknowledgments

Research for this book was funded by the Australian Research Council through a Discovery Grant (2013–2016), "From alchemist's den to science city: architecture and the expression of experimental science." We are deeply grateful to the anonymous reviewers who saw value in our original proposal.

While the ARC's funding made it possible for us to visit bioscience research centers across the world, those visits were made worthwhile because of the generosity of architects, facility managers and scientists, and the organizational skills of their management teams. We would like to thank especially:

India Wallace, Michelle Gheorghiu, and Sharon MacGowan from SAHMRI and Georgia Singleton, Woods Bagot;

Robert Friedman, J. Craig Venter Institute, and Ted Hyman, Zimmer Gunsul Frasca;

Randal Jones, Cold Spring Harbor Laboratory, and Jim Childress and Todd Andrews, Centerbrook Architects and Planners;

Jamal Kurtu, Pharmaceutical Sciences Building, University of British Columbia, and Roger Hughes, Hughes Condon Marler Architecture and Design;

Reimund Fickert, PRBB, and Manuel Brunet and Nil Brunet, Brullet de Luna Arquitectes;

Jesús Purroy and Miriam Martínez Vivo, Parc Científic de Barcelona;

Warwick Murphy, Peter MacCallum Cancer Centre, and Christon Batey-Smith and Rohan Wilson, DesignInc;

Eva Fernández, Antonio Vaíllo, and Juan Luis Irigaray Bascaran, Vaillo+Irigaray, and David Escors Murugarren and the Miguel Servet Foundation at NavarraBioMed;

Chris Pelling, Jeff P. Maskell, and Maria Caballero-Blaya, Blizard Institute, Queen Mary University of London;

Ralf and Carla Streckwall, Max-Delbrück-Centrum Für Molekulare Medizin, Berlin;

Stéphane Piquand and Sophie Artero Genzyme Polyclonals, Lyon;
Maria João Villas-Boas, Fundação Champalimaud, Centre for the Unknown, Lisbon.

Periods of focused research over the course of the grant were made possible by the
University of Sydney and the University of Queensland. We also had the opportunity to present our ideas to: Institute for Advanced Studies in the Humanities and
the Edinburgh School of Architecture and Landscape Architecture at the University of
Edinburgh; the Institute of Health and Biomedical Innovation, Queensland University
of Technology; Architecture Theory Department, TU Delft; Taubman College, The University of Michigan, and; the AEDES Network in Berlin.

We would like to thank the anonymous reviewers of earlier papers on biosciences
and architecture that were published in the *Architecture Research Quarterly*, the proceedings of the Society of Architectural Historians of Australian and New Zealand, *Industries
of Architecture*, and *le Journal Spéciale'Z*. Over the past four years we have benefited
from conversations with Mark Dorrian, Amy Kulper, Tilo Amhoff, Katie Lloyd-Thomas,
Mathew Aitchison, Hélène Frichot, Naomi Stead, Juliet Odgers, Stephen Kite, Silvia
Micheli, and Andrew Holder.

We were greatly helped by the contribution and assistance of Dr. Russell Hughes and
Dr. Alison Holland, and Sean Bryen and Malakai Smith. Dr Hughes's research on New
York's biosciences ecosystem deserves special mention. Analytical illustrations of laboratory buildings were drawn by Quoc Anh Ho, Aiden Morris, and Carlota Marijuan-
Rodriguez. While we visited each of the building case studies, our own photographs
suffered from being taken while touring the building at a smart pace or in poor light.
We are thus grateful to the archives, architectural practices, and professional photographers who helped fill in the gaps with beautiful images. A special mention, however,
must be made of the amateur photographers, many of them eminent scientists for
whom photography is not their core business, who have contributed to the book.

Lastly, we led design studio and theory courses at our respective universities that
were informed by our research, and we thank the students whose enthusiasm for the
subject reinforced our own. We hope students who are compelled to read this book will
find it enjoyable.

1 Introduction: Cathedrals of Science

The laboratory building is as significant to the twenty-first century as the cathedral was to the thirteenth and fourteenth centuries. Scientific knowledge has largely displaced theology and philosophical speculation as a framework for understanding life and the universe.[1] The privileged status of science in knowledge formation is reflected in campaigns by universities, governments, corporations, and developers to construct scientific research buildings and precincts. In this book we focus on the architecture of biosciences research. The biosciences are concerned with the biological aspects of living organisms, with life itself. Of all the collective human projects aimed at new knowledge, it is this field whose discoveries seem to bear most intensely on our lives and the future of our planet.

Yet it would be naïve to think that investment in this area is driven by curiosity and altruism alone, for biosciences research is inseparable from its application in biotechnology—including the pharmaceutical industries. In 2017, the revenue of the worldwide pharmaceutical market was USD1,143 billion, having doubled in just over a decade.[2] Current biosciences research projects are aimed at extending human life, curing cancer, reversing climate change and species extinction, replacing fossil fuels, transforming pollutants into benign substances, feeding the world's growing population, and rewriting the genomic make-up of living organisms. It goes without saying that the stakes are high and the rhetoric around the field quixotic. Public and private investment has been bullish, while the expectation of profit bears little relationship to the slow pace and unpredictable outcomes of research and its translation into pharmaceutical product, clinical procedure, or environmental intervention.

Not surprisingly, investment in the biosciences has delivered an extraordinary building boom since the coming of the new millennium. But there is more to the architectural expression of the biosciences than accommodating the people, animals, services, and equipment needed for experimentation. The very *act* of committing vast resources into permanent structures is a declaration of faith in science and its potential discoveries. Budgets are sufficiently generous for these buildings to fulfill the particular ambitions and conceits of architects, clients, and occupants—indeed, for even the most preposterous ideas and luxurious fantasies to be realized. We might learn much about the biosciences from architecture, for here they are given their most legible, insistent, and nuanced articulation. In laboratory buildings for the biosciences we can learn much about architecture, too: its potential, its expressive force, its social effects, its disciplinary assumptions. Architecture is a symptom of the forces that work through and upon it. Architecture also structures and produces desire and plays a part in the disciplining of subjectivities. In laboratory buildings, architecture is all this, and more.

Many, however, still imagine laboratory buildings as "warrens of endless corridors and closed doors behind which new discoveries are squirreled away"[3] and an "unloved type where boffins are thought never to look up from their experiments or talk to one another."[4] The leap from "anonymous prefab shed, housing a functional stack of labs and pipes" to today's architectural edifices is overdrawn, and these claims need to be understood in light of the history of the laboratory building. The period of anonymity and pragmatism turns out to have been relatively short, as well as being much exaggerated for the purposes of dramatic contrast with today's buildings. The history of laboratory architecture begins with a new form of knowledge emerging in the seventeenth century, which privileges the repeatable and observable experiment and assumes the availability of nature for transformation. To ensure repeatability, experimental knowledge required a neutral, equipped, and protected space that excluded anything (and anybody) outside of the parameters of the experiment. Such a space was not invented from scratch. The laboratory as an artifact—as a stand-alone and purpose-built structure—brought together a set of extant spaces and activities.[5] The furnace rooms of alchemical transformation are often cited as the laboratory's precursor, and they linger in the fantasy of the scientist secreted in the basement. Equally important for the laboratories of the life sciences are the sites and practices of acquisition and display, such as the *Wunderkammer*. These cabinets of curiosities brought together the encyclopedic collections of sixteenth-century gentlemen—from fossils to works of art, the stranger the better. The impulse to collect led to the formation of the museum in the eighteenth and nineteenth centuries, which itself developed as a key site for pursuing and exhibiting scientific knowledge. The places in which plants and animals could be cultivated outside of their natural habitats, such as the glasshouse, the zoological garden, and the aquarium, are also ancestors of the laboratory. The traces of these can be seen in the growth rooms and cabinets, the pathogen-free colonies and vivaria of contemporary research buildings. Early science was also taking place in hospitals and

in asylums for the insane. Rooms designed for observation, such as the anatomical theater and the planetary observatory, as well as places set aside for the discussion of philosophical ideas, are also precedents. After the seventeenth century, these included "the
public rooms of the residences occupied by public persons"[6] and later clubs, salons, and
learned societies established outside the home. Lastly, laboratory buildings incorporate
spaces for contemplation and dissemination that derive from the library, the scholar's
study, and the classroom.

Bruno Latour argues that by the nineteenth century, laboratories were "becoming
special places to make specific goods—in this case facts—for a new emerging market, the
scientific market."[7] The laboratory was to science, what the factory was to capitalism.[8]
The standardization of experimental protocols and materials corresponds to the mass
mobilization of scientific instruments and sees the expansion of new experimental set-
ups.[9] By the mid-twentieth century the laboratory had brought transformation, collection, cultivation, observation, discussion, contemplation, and production together in
a recognizable format. The period coincided with the maturity of the modernist style.
Modernist principles yielded open-plan laboratories organized by repeated modules of
structure and furnishing, served by separate offices and service zones.[10] Scientific ideals
of rationality and fitness for purpose intersected with the functionalist strands of architectural modernism. So closely aligned are modernist functionalism and the laboratory
program that it became possible to forget the presence of architecture.[11]

Representative of this period is the Hoffman Laboratory of Experimental Geology
(1963) at Harvard University by Walter Gropius's The Architects Collaborative (TAC)
(figure 1.1). One can see numerous lesser-known examples by lesser-known architects, which employ the same planning conventions—the J. Walter Wilson Laboratory
(1962) at Brown University by Robinson, Green, and Beretta, for example, where each
floor is dedicated to a different branch of biology, from animal physiology on the first
to plants on the fourth. Even more severe than these two is the Richard King Mellon
Hall of Science (1968), designed by Mies van der Rohe for Duquesne University, Pitts-
burgh. With its façade of black panels and small dark glass windows, it is unwelcom-
ing and introverted, yet was awarded "Laboratory of the Year" in 1969 by *Industrial
Research Magazine*. Jose Luis Sert's muscular Science Center for Harvard University
(1972) has a stepped form to accommodate outdoor patios, upon which scientists
were to socialize on the rare occasions that the weather would permit (and is, vaguely,
suggestive of the polaroid camera invented by the building's primary donor, Edwin
Land). These buildings tended to be constructed and owned by a university, buried
deep in the campus, and narrow in their disciplinary attachments. Architecturally, each
is notable for its focus on the functional demands of its wet laboratories, its visual

Figure 1.1
Hoffman Laboratory, Harvard University, Cambridge, Massachusetts (1960). Architects: The Architects Collaborative (TAC), led by Walter Gropius. Photograph by Daderot.

impermeability, its lack of social amenity for its occupants, and its dearth of facilities for public engagement.

The modernist laboratory captures a broader conceit: namely, the separation of science and society. This separation reached its climax in the decades of the Cold War. This is when the alliance of scientists and militaries transformed the confinement associated with laboratory buildings into geographic isolation and architectural fortification. Remote places became the sites of science. The Los Alamos National Laboratory (1943) in New Mexico is a prime example, Ozorsk (now Ozyorsk) its Soviet counterpart. Indeed, the "closed cities" for scientific and military research in the Soviet Union could be seen as the pinnacle of this expression of isolation, with all their denizens, scientists and otherwise, effectively quarantined. In the United States, a notable example of defensive architectural expression is Paul Rudolph's Burroughs-Wellcome Laboratories (1969) in North Carolina (figure 1.2). The building is not geographically remote, but Rudolph's design for the pharmaceutical company's laboratories and headquarters is dramatically fortress-like, resembling a bunker even, with jutting angles and exposed

Figure 1.2
Burroughs-Wellcome Headquarters and Laboratories, Durham, North Carolina (1969). Architect:
Paul Rudolph. Courtesy of the Paul Rudolph Foundation.

aggregate concrete walls inside and out. According to the Paul Rudolph Foundation,
the local community still refers to it, as "the spaceship building."[12]

While entwined in the politics of the time, the midcentury laboratory was constructed in such a way that it *conceptually* excluded the public, along with the sociality of scientists and the dynamics of competing teams and disciplines. Louis Kahn's
design of the Salk Institute for Biological Studies (1965) in La Jolla, California, is one
of the first to register the changing desires and conditions of science. This building's
importance in architecture and the social sciences is such that we devote the following
chapter to its influence. What Kahn and the scientist Jonas Salk recognized was that
efficiency and isolation were not the only driving forces of scientific practice, and nor
were such forces the key determinants of the laboratory's architectural expression. The
shift they recognized, as Bruno Latour explains, derives from an explicit entanglement
of the scientific and the social. He has suggested that a culture of "science" came to
be replaced by a culture of "research," which engages ideology and emotion. The shift
reflects broader changes that swept across many disciplines in the wake of world wars

and cold wars. The distinction for Latour is that science is "cold, straight and detached; research is warm, involving and risky."[13] Latour sums up the situation as "a science freed from the politics of doing away with politics."[14] In some ways, the shift heralded a return of the theatrical public demonstrations of science that characterized the mid-nineteenth century, not so much in live performances, but through a range of media, including architecture.[15]

Beyond Functionality

The explicit entanglement of science and sociopolitical concerns has seen the scientist reemerge onto the civic stage with a broad range of skills, including those related to fund-raising, team-building, educational outreach, media wrangling, community liaison, and public relations. To enable this wider remit, new buildings for the biosciences go well beyond the functionality of their technical apparatus. Take, for example, the Cha Bio Complex (2015) in Pangyo, Korea, designed by the American firm KMD Architects and Korean architects Designcamp Moon Park. Alongside its research laboratories, vivarium, offices, and tissue bank are conference, exhibition, and cultural facilities including a 3D media hall, dormitories for researchers and guest housing for important visiting researchers, offices for venture capital businesses, an invitro fertilization clinic, a health club with lap pool, a cafeteria and cafés, a library, meadow gardens, and, to top it off, a lounge with a "medieval-style pipe organ."[16] Semipublic spaces for socialization, collaboration, presentation, and persuasion now take up a greater proportion of habitable floor space than wet laboratories and clean rooms. In addition, there are dramatic atria, sculptural stairs, and circulation routes incorporating conversation nooks and views into the laboratories. No longer hidden away, bioscience laboratories now occupy prime real estate, perhaps none more so than the Parc de Recerca de Barcelona (PRBB), where its scientists play volleyball on the Platja de la Barceloneta (figure 1.3).

The novelty of the bioscience laboratory building today is that architecture is regarded as a means to achieve organizational aims. Architecture is engaged to recruit star scientists, inspire young people to embark upon careers in science, and attract philanthropy and industry partners. It is architecture that sponsors conversation and collaboration between scientists and enhances the performance and health of employees, and, thus, it accelerates discovery. Architecture is charged with engaging the public in scientific endeavors. It does so literally through spaces in which events for advancement and industry participation might be hosted, as well as by communicating the importance (and content) of scientific research. Nigel Thrift describes the new buildings that are part of the biosciences boom as "traps for innovation and invention."[17]

Figure 1.3
Barcelona Biomedical Research Park (PRBB) (2006), Barcelona, Spain, from the sea. Architect: Manuel Brullet and Albert de Pineda. Photograph by Javier Ortega Figueiral (2012).

These ambitions respond to a diverse array of ideological and economic forces encompassing neoliberal concepts of knowledge work, the reification of innovation, and the free-market entrepreneur. They also respond to technological change and shifting financial institutions and structures. The corporatization of universities and the concomitant uptake of collegial organization in industry are factors, as is the growth of speculative real-estate development in the provision of science facilities. Belief in science as a means of addressing aging (indeed, mortality itself) and other "big" problems also fuels the growth in biosciences research. Every nation-state from Botswana to Bolivia has official policies for the promotion of biosciences research and development.[18] Laboratories have become sites of power and prestige, faith and ambition. They are places that link nations into the global trade of scientific knowledge.

The contemporary research laboratory has joined the museum and the art gallery as a building type deserving of a prominent site and a signature architect, iconic treatment and civic pride, public investment and private philanthropy. The Lewis Thomas Laboratory for Molecular Biology (1986) at Princeton University by Venturi, Scott Brown, and Associates is a harbinger of this change. The architects added a patterned brick skin and classical details to a functional laboratory layout by Payette Associates. Although Venturi, Scott Brown, and Associates were already well known, it was another five years before Robert Venturi (without Denise Scott Brown) was named a Laureate of the Pritzker Architecture Prize, the highest accolade in the profession. The more recent run of laboratory commissions to Pritzker Laureates has come *after* their win: clients are clearly concerned with the prestige that an awarded architect brings to the building. Indeed, on the Cornell University homepage for Weill Hall (2008), the opening sentence declares the building "was designed by Pritzker Prize–winning architect Richard Meier."[19] Meier won in 1984. His design for the Ithaca campus, constructed at a cost of USD 162 million, is characteristically white, crisp, and corporate. Other "starchitects" have made better use of the opportunity to deploy their form-making and expressive prowess beyond the application of surface patterns. At a cost of USD146 million in 2003, the curvaceous James H. Clark Center for Bio-X at Stanford University was designed by architect Sir Norman Foster, winner in 1999 (figure 1.4). Pritzker Laureates, Frank Gehry (1989), Rafael Moneo (1996), and Zaha Hadid (2004) are respectively responsible for: the Ray and Maria Stata Center at MIT (2004; USD 283.5 million), the Laboratory for Integrated Science and Engineering at Harvard University (2007; USD 155 million), and Biopolis in Singapore (2003–2006; USD 457 million). In the following decade, Jean Nouvel (Pritzker 2008) completed the Institut des maladies génétiques Imagine in Paris (2014; USD 46 million), and Renzo Piano (Pritzker 1998) the Jerome L. Greene Science Center in New York (2016; USD 250 million).

Figure 1.4
James H. Clark Center (2003), Stanford University, Stanford, California, at night. Architect: Foster and Partners. Photograph by Anirudh Rao (2010).

Pritzker Laureates Rafael Moneo, Tadao Ando, and Fumihiko Maki, as well as Alvaro Siza, Kazuo Sejima, and Ryue Nishizawa of SANAA have all designed laboratory buildings for the Basel campus of the pharmaceutical giant Novartis. Herzog and de Meuron (2001 Laureates) completed an office tower, Asklepios 8 (2015), at Novartis, as well as the headquarters (2010) and laboratory (2013) for rival pharmaceutical company, Actelion, in Basel. Biosciences architecture is now divided between the laboratory planning consultancies and specialists, who ensure the functionality of the wet laboratory proper, and the high-profile design architects commissioned to design the surrounding spaces of research buildings and express the conversational and convivial aspirations of the scientific organization.

Cathedral-Like

Given their cultural and economic significance, along with the role of scientific knowledge as a secular belief-system, laboratories are often likened to cathedrals. An article in *Nature* on the Francis Crick Institute (2016), London, was titled "Europe's Superlab: Sir Paul's Cathedral."[20] Sir Paul Nurse, diplomatically avoiding the implied slight that the building is a vehicle for his personal aggrandizement, responded, "You need to go into a building and feel inspiration. That is what is so beautiful about a medieval cathedral— you are inspired whatever your religious beliefs might be."[21] The architects, HOK, have encouraged the analogy, describing its entrance as "cathedral-like" and designing the building around a lofty nave.[22] The Francis Crick Institute cost USD 790 million (GBP 650 million) to construct, a considerable outlay at a time of economic austerity, which, like a cathedral, required the indulgence of the public and the promise that future benefits would accrue—in the form of a longer life, rather than an afterlife. At nearly 1 million square feet (93,000 square meters), it has just slightly less useable floor area than the British Museum. This British example is not exceptional in the scale of its resource investment, although it is unusually pedestrian in its architecture.[23] The Belfer Building in New York City cost USD 650 million, while Australia's Victorian Comprehensive Cancer Centre (2016) was constructed for USD 761 million (AUD 1 billion). But it is not only the largest laboratory projects that have social, affective, and aesthetic aspirations.

Sociologist Karin Knorr Cetina also engages the analogy between the cathedral and the laboratory. For her "the laboratory houses within itself the circuits of observation and the traffic of experience which twelfth- and thirteenth-century cathedral builders brought about through travel, and it includes an exchange of specimens, tools, and materials."[24] Knorr Cetina is interested in the social construction of scientific knowledge, which she sees as being enabled by laboratories because of their place in a wider network of experimentation and knowledge exchange. And yet, there is a profound difference between the laboratory and the cathedral, which gets passed over in this analogy. For those building them, the cathedral was also the *subject* of aesthetic and structural experimentation. Going higher and leaner was the experiment. We will argue that in the contemporary laboratory, *scientists* are the subjects in an experiment in social engineering and public persuasion, for which there is a dearth of evidence and a vague methodology. Through architecture, desired behaviors are to be elicited. The concept, known as "architectural determinism," has been largely debunked in academic circles, but is alive and well in the design of laboratory buildings. In the case of hospitals and schools, it has morphed into "evidence-based design." The tenacity with which architects have held on to architectural determinism, as Robert Gutman and Barbara Westergaard skeptically pointed out in 1974, has a lot do with the individual architect's need to justify his or her work.[25] It is not so much cynical opportunism that pervades the laboratory design sector, but a feverish optimism that both architects and scientists are caught up in about their professional responsibility, and ability, to improve the world. This makes laboratory buildings sites of intense rhetoric and hyperbole, as well as of formal and spatial experimentation that spans the gamut from the banal to the brilliant. Their recurring features and effects will be interrogated in the following chapters.

Architect Frank Gehry, another Pritzker Laureate, reveals both the desire for and the lingering doubt about architecture's capacity to shape scientific discovery. When speaking of his design for the Ray and Maria Stata Center (2004) at MIT, he hopes "accidental or contrived" collisions between people "will lead to the breakthroughs and the positive results." Gehry says, "I think that's really going to work."[26] He speculates, but does not know. Scientists would never excuse their peers for such imprecision, but when it comes to architectural speculation, they may not be so exacting. Rodney Brooks, the roboticist and director of the Stata Center, discussed with *Wired* magazine his research group's relocation to the Gehry designed laboratory. He said, "Maybe it will destroy us. Who knows? I prefer to be optimistic."[27] This book is not concerned with whether or not new laboratory buildings perform as expected. It was never our intent to undertake post-occupancy evaluations, and, in any case, it is impossible to ascribe discoveries to architecture rather than to, say, funding, effective leadership, timing, good fortune, or genius. Even where the co-location of previously dispersed researchers is found to have had a positive effect on research activity and collaboration, this is not necessarily tied to specific architectural qualities or tactics. Indeed, transdisciplinary co-location was a feature of MIT's Building 20 (1943–1999), the hastily erected "temporary" structure from which many of the occupants of the Stata Center had decamped. The degraded and undesirable qualities of Building 20, not any architectural merit, made it a place where various innovators, from Noam Chomsky to the first hackers, could work creatively.[28] With this in mind, *LabOratory* is concerned with the nature and consequences of the expectations scientists and architects have for their new buildings, not the elusive evidence that might justify them. We have identified three common and consistently repeated ambitions. These are easier to understand by first returning briefly to the cathedral.

A Common Rhetoric

The cathedral requires three things of its architecture. First, it secures a protected, or sacred, space set apart from the profane. Second, its architecture (and embedded art) expresses the doctrine and teachings of the faith in ways that reach an illiterate audience. Third, it choreographs the performance of the liturgy and, thus, the relationships the members of the congregation have with each other, with their church and clergy, and with their Christian God. The first demand is fulfilled by the detachment of the cathedral from other buildings. While usually centrally located in an urban setting that it anchors and dominates, the cathedral's separation from the world of commerce is emphasized by an empty space or plaza between it and the town hall or market. Separation is further reinforced by an elaborate and over-sized portal, and by the fineness of its stone masonry. Its dark, cool, and quiet interior heightens the contrast with the outside world. The stained-glass windows, carved gargoyles, statuary, and other narrative and iconographic works of art are expressions related to the second demand a cathedral makes of its architecture. The spatial unity of the composition, the processional aisles and ambulatory paths for ritualized events, the shared pews, and the location of the clergy at the altar relate to the third objective. The cruciform plan addresses both representative content and spatial choreography by reminding the congregation of Christ's sacrifice and catering to procession and formality. The verticality that organizes structure, surface, and space—from the elongation of the ribbed vaults to the towers flanking the entrance—expresses the idea of an all-seeing God above the earth. While historians of medieval architecture would most certainly contest this simplification and argue the many differences between, say, the German, English, and French Gothic cathedral, there is nevertheless, a recognizable formation or typology.

In the eighteenth century, the French architect Julien-David Le Roy used the consistency of the Christian church plan to mount an argument for the basis of architectural beauty in the timeless principles of human aesthetic perception. The single plate in his *Histoire de la disposition et des forms différentes que les Chrétiens ont données leurs temples* (1764) shows plans of churches built between AD 320 and 1764 in different places (figure 1.5). If there was any historical development in the type, it is not clear, for they are not arranged on the page in any sequence. Minor variations and details in their design are textually and graphically suppressed. LeRoy—in response to criticism—attempted in the revised edition of *Les Ruines des plus beaux monuments de la Gréce* (1770) a new drawing that brought together temples and cathedrals from different cultures in chronological order. But, as Jeanne Kisacky attests, the ambition remained "to uncover universal laws structuring specific forms and events," much as his three scientist brothers were doing in their respective fields of horology, medicine, and electricity.[29] Le Roy was one of the first architectural historians to adopt from natural history the taxonomical approach to drawing subjects at the same scale, side by side, according to their type. J. N. L. Durand developed the genre further in his *Recueil et parallele des édifices de tout genre, anciens et modernes* (1801), a pictorial atlas of buildings sorted either by purpose (hospitals, cathedrals, prisons) or by architect (Palladio, Inigo Jones).

In the nineteenth century the French architect and theorist Eugène Viollet-le-Duc reimagined the Gothic in the process of restoring cathedrals across France. Drawing upon biology, Viollet-le-Duc formulated a *cathédrale idéale*, which was intended to reveal the rational structural principles common to all built cathedrals. Real cathedrals were viewed as variations on, or transformations of, this theoretical type. For him, all of the elements of the cathedral, including its decorative and iconographical design, grew from the functional necessity of satisfying the ambition for height. Viollet-le-Duc's belief that successive generations of cathedral builders were striving toward the perfect resolution of the type reflects an evolutionary idea. The architecture of the cathedral "develops and progresses as nature does in the creation of beings; starting from a very simple principle which it then modifies, which it perfects, which it makes more complicated, but without ever destroying the original essence."[30]

Compared with the cathedral, there seems to be little foundation for grouping laboratory buildings together and no possibility of drawing a *laboratoire idéale*. Formally, they are as irreconcilable as a transistor radio is to a drone, a talking doll, a cordless shaver, or an electric bicycle—although all are battery operated and thus have similar inner workings. The architecture of the contemporary biosciences laboratory is as heterogeneous as these gadgets, despite each having a wet laboratory as its core purpose and inner organ. The accounts architects, clients and critics give of laboratory buildings, however, are strikingly homogenous. The following description of the biochemistry laboratory at the University of Oxford (2008) designed by Hawkins Brown is representative. We are told that the building

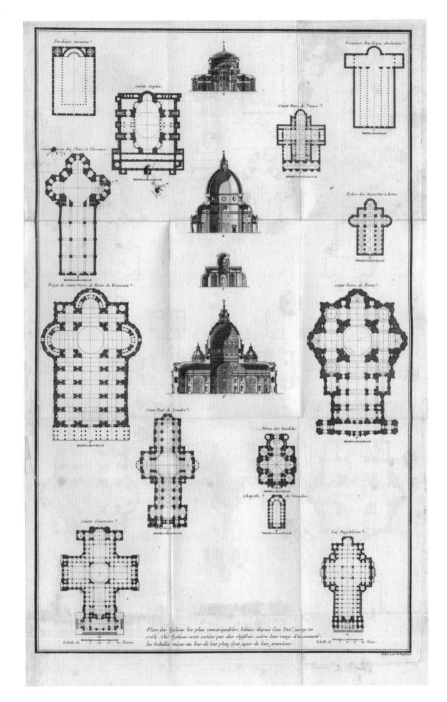

Figure 1.5
Julien-David Le Roy's Genealogy of the Christian Church, 1764, from *Histoire de la disposition et des formes différentes que les Chrétiens ont données à leurs temples* (Paris; Desaint & Saillant, 1764). Courtesy of the Bibliotheque nationale de France.

ensures the 300 researchers working there communicate as much as possible. The traditional layout is reversed here, labs are on the outside, divided by clear glass walls from the write-up areas, which are open to the vast, five-story atrium. Everyone is visible. Open staircases clad in warm wood fly across the atrium at odd angles, and each floor hosts a cluster of inviting squashy leather chairs and coffee tables, giving the impression of an upmarket hotel.[31]

The *Whole Building Design Guide* is published by the National Institute of Building Sciences in the United States and is far less effusive. Most of its attention is given to pragmatic questions of servicing and planning, yet it gives this advice to the architect of research laboratories:

> Entrances and public greeting spaces must make the first impression unforgettable. A mix of scientific displays, interactive flat panel screens and real-time or digital video views into best teaching and research labs in action should be a basic requirement. The design should provide an unlimited access to the rich world of discovery.[32]

Every laboratory architect, every laboratory client, declares the same three aspirations: to eliminate boundaries; to communicate the potential benefits and interest of its research programs; and to foster collaboration amongst its scientists. In brief, these three rhetorical deferrals will be elaborated across this book in corresponding key sections: "Boundaries," "Expression," and "Socialization."

Firstly, *boundaries*. "Despite various characterizations of science as 'public knowledge,'" writes Steven Shapin, "it is made and evaluated in some of our most private places.... You do not wander into CERN or SLAC. We typically now enter the places where scientific knowledge is made only by special arrangement and on a special basis: we come as visitors, as guests in a house were nobody lives."[33] Biological containment and the securitization of costly experiments and equipment would seem to justify the attention given to boundary-making and policing, yet the concern for laboratory boundaries lies equally with the need to legitimize scientific expertise, to create a place of order apart from the chaos of the world. These concerns are rarely addressed explicitly. Instead, claims are made about the *openness* of the laboratory, its accessibility and *transparency*. The tension between containment and isolation and the demand for institutional accountability animates the design of the contemporary laboratory. It is a tension that is particularly acute at successive thresholds and openings in these thresholds: the perimeters of laboratory grounds and compounds; the envelope of buildings; and the walls around wet labs and clean rooms.

Secondly, *expression*. Nigel Thrift notes that the new generation of biosciences buildings "often include an explicit attempt to represent 'life', whether that be swooping architecture, some forms of public display of science or similar devices."[34] More specifically, attempts are made to convey scientific content or subjects. Of the Blizard Building (2005) at Queen Mary University of London—a building we discuss in detail in chapter 6—architect Will Alsop proposes that "the very fabric of the building speaks about science."[35] Replacing ignorance and fear of science with desire and understanding is assumed to lead to greater financial and community support for experimental research (or at least less resistance to it). Architecture is expected to assist in the positive representation of scientific content.

The scientific content drawn from the life sciences tends to be living organisms and their composite parts. In fact, it includes the technical images with which science represents those subjects and is able to make them work for science—the double-helix of DNA, the periodic table, and the chromosomal map, to name a few. The modes of communication of scientific content in the building type are extraordinarily diverse and stretch from analogies that are obvious, such as the brain-shaped neurosciences research center or the application of images of cells on a façade, to analogies that are ambiguous and multivalent. For example, the colored Rorschach inkblot patterns laminated to the glazed front elevation of the University of Oxford's Biochemistry Building (2008) are intended as reminders of the suggestive nature and imprecision of perception—a critique of scientific truths that may itself get lost in translation.

Scientific content also includes the values and identities of individual scientists, and expressions of this may be as unambiguous as naming a building after an acclaimed researcher or entail more complex strategies of portrayal and allegiance. Some laboratories come close to the use of gargoyles and depictions of liturgy and saints in stained glass. Not surprisingly, problems of translation and transformation in this situation abound. The convergence of architecture and science in the design of laboratory buildings is further complicated by the traffic of concepts, techniques, and images between the two fields. For example, double-helical stairs, or *escalier Leonardesca*, predate Watson and Crick's 1953 announcement of the discovery of the structure of the DNA. Their contemporary use in science buildings may pay homage to Leonardo da Vinci, but the iconic status of DNA in popular culture overwhelms their architectural ancestry.[36]

Thirdly, *socialization*. Laboratory design is said to accelerate discovery by provoking serendipitous conversations. Typically, the means of enhancing socialization has been conceived by the architects as the provision of attractive shared areas and circulation spaces that encourage scientists to bump into each other. The point of the café, the open stair, or squashy leather chairs is not merely to make it *possible* for scientists to socialize, or even to *entice* them, but to convey the desirability of (and potential reward for) social behaviors in the workplace. We will discuss in chapter 12 how the

Figure 1.6a
View of the interior of The Advanced Training Centre (2010) of the European Molecular Biology Laboratory (EMBL), Heidelberg, Germany, with its double-helical ramps. Photograph by Jun Seita (2014).

Figure 1.6b
The exterior of the EMBL is much less dramatic than its interior. Photograph by authors (2015).

key idea of socializing slides into subject production, for these devices are engaged in the disciplining of the scientist subject within a neoliberal frame of immaterial work. We do not mean that scientists are passively manipulated. Many scientists subscribe to an idea of their work as creative, social, and independent of management strictures. Some are involved in the design briefing and review process, where they reiterate their hope for spaces that foster the workplace culture they desire. The socialization of the scientist is a cultivation of behaviors and concretization of relationships, teams, and hierarchies. At the same time, it is a representation of those desired behaviors and relations.

If, the second ambition sees architects focusing on *science as content* for expression, in meeting the third aspiration, the focus is on the *scientist* as content producer and knowledge worker. There is, as with the cathedral's crucifix plan, a grey area where spatial choreography and expression converge. The need to reassure the public that scientific work is collaborative and that clandestine projects are impossible directs architecture toward the representation of interaction, inclusivity, and transparency. Hence, some laboratories are designed to convey collision through dynamic forms or to express community through the inclusion of spaces intended to resemble village squares and streets. The double-helical plan of the Advanced Training Center (2010) on the Heidelberg campus of the European Molecular Biology Laboratory (EMBL), collapses scientific imagery with socialization ambitions. Its disorienting and protracted circulation ramps force its occupants to circumnavigate their way past occupied offices and each other, while its exterior form struggles to convey its conceptual referent (figure 1.6).

While we cannot make clean divisions between the laboratory's boundary modulation, its expressive strategies, and its social organizations, this book has been loosely organized around these three themes and in that order. The first bears upon architecture as the creation of discrete worlds, as a means of insulating or isolating the places in which scientific knowledge is made, while at the same time allowing controlled trade and movement. The second concerns architecture's communicative capacity and the metaphors, images, and ideas that architecture claims to draw from science. The third foregrounds the political and social forces at work in, and on, scientific organizations, the construction of the scientist subject, and architecture's role in reproducing, or shifting, social structures and conventional behaviors. But our purpose is not simply to identify features and create taxonomies.

At the heart of the book are several questions: *Why are today's laboratories so radically different from their modernist predecessors and from each other? How do architectural differences between laboratories emerge? What are the effects of these differences?* These questions are not unlike those faced by Charles Darwin in contemplating the myriad forms of animal and plant life. The puzzle demands more than, say, ascribing the turtle's shell to its need for protection. Linking form to function does not explain how, or why, the turtle took this route rather than the articulated carapace of the lobster or the aardvark. In other words, it requires that we ask about the mechanisms and forces of change in the entire biosciences ecosystem. Darwin's image for the forces of change was "a thousand wedges," and his image for the ecosystem in which difference is played out is "the entangled bank." We need to view each laboratory in *relational terms*, alongside other laboratories from which it departs or whose characteristics it shares. We cannot ascribe their variations to a single narrative, as did Viollet-le-Duc when explaining the development of the cathedral in terms of the inherited knowledge of its builders. We need to understand the laboratory's adaptation to, and fitness for, the complex environments in which it competes for scientists, funds, and industry partners.

The Flowering of Diversity

When we commenced this project, we collected hundreds of photographs and architectural drawings of biosciences laboratories that have been built since the completion of the James H. Clark Center for Stanford University's Bio-X (2003). Each has, as its core anatomy, one or more wet laboratories that are securely bounded (with one exception) and easily identified in a plan. We focused on buildings with the most common classification for biosciences research at Biosafety Level 2. (Biological containment is ranked from levels from 1 to 4. Laboratories at Level 4 are designed to contain hazards—even in the event of nuclear war—for which there is no known cure.) Level 2 meant we could visit each of the buildings discussed in this book. The interiors of wet laboratories vary little. This is due to the constraints that arise from the observance of experimental protocols, complex service requirements, the need for even and bright artificial light and impermeable and anticorrosive finishes, and the standardization of equipment and fixtures. It is the comparative stasis of the bench space for experimentation that has seen laboratories described as "all the same (more or less)" and as constituting "a universal cultural space."[37]

Massing them together, we noted some curious and arbitrarily recurring features such as the inclusion of sculptural double helical and helical stairs and the use of a lime green accent color in write-up and social spaces. But beyond a few rhetorical elements and the enclosure of the wet laboratory space, the architecture of contemporary laboratory buildings is extraordinarily diverse and idiosyncratic. Indeed, they are as different to each other as a zebrafish is to a tiger, although both are striped and both are vertebrates.

The differences between the sites of biosciences research are not just skin deep, as the diagrams in figure 1.7 and figure 1.8 make very clear. In figure 1.7 the configuration of Physical Containment Level 2 PC2 wet laboratory spaces, in blue, is shown within the larger building volume. In figure 1.8, the primary vertical and horizontal circulation routes are shown in brown. Some, like Weill Cornell's Belfer Building (2014), designed by Ennead Architects, and the Alexandria Life Sciences Center (2014) by RMJM are midrise towers in New York City, indistinguishable from the corporate office buildings around them. Others are dispersed and resemble quaint domestic buildings or modest civic institutions. They vary a great deal in size, from a couple of rooms to major pieces of civic infrastructure. They can be found in condensed urban biosciences precincts or, amongst resorts on a stretch of Mediterranean coastline, as is the PRBB (2008). Internally, the organization of wet laboratories varies widely from being stacked vertically, ranged in a horizontal line or radially, located below or high above ground, on display or deeply buried behind other spaces. Some incorporate animal houses (vivaria), while others do not. Some have residential facilities, although most do not. All have cellular and open-plan offices, but in different configurations with varying degrees of adjacency and visibility in relation to their wet laboratories and public spaces.

In accounting for the formal diversity of laboratory buildings, we have situated architecture within the very rich and broad ecology of scientific rhetoric and persuasion. We emphasize that architecture is a conglomeration of elements: architects, builders, clients; contracts, budgets, subsidies, and fees; organizational structures and missions; programs of use and projects envisioned; building and planning regulations; equipment and servicing; sites and neighborhoods; and structure, construction materials, proprietary fixtures, and material. These weigh on the design of each laboratory to different degrees. Some laboratories are influenced by the architectural preferences and personalities of eminent and powerful scientists and clients; others are almost uncompromised distillations of the architect's signature approach to form and material. Some have generous budgets and benefit from a skilled local construction industry; for others the opposite is true. Many are concerned with transparency and public engagement, while others are focused on amenities for the scientists within. We have not made our selection based on architectural excellence or with the aim of representing all approaches to the building type. This is neither a comprehensive survey nor a comparative analysis. Instead, we have identified buildings that have rich and complex, even messy, stories to tell: laboratories through which we can articulate some of the complex mechanisms at stake. We fixate upon successive and singular elements in each of our case study chapters not in order to reduce, or simplify, but to arrive at a

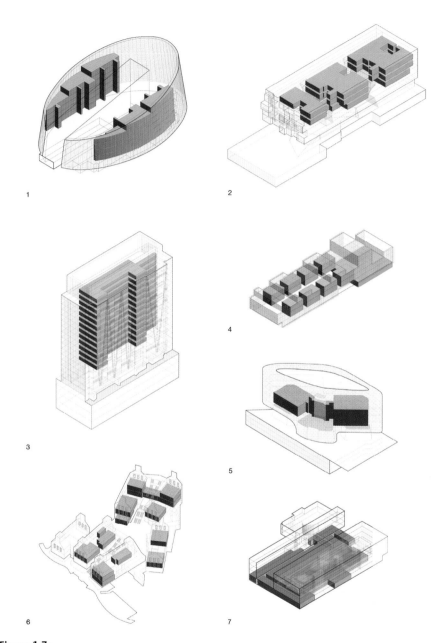

Figure 1.7a
Biosciences research buildings exhibit no formal consistency, as evident in the diagrams showing wet laboratories in blue. 1. Barcelona Biomedical Research Park (PRBB), Barcelona. 2. Pharmaceutical Sciences Building (PSB), Vancouver. 3. Belfer Building, New York. 4. Navarra BioMed, Pamplona. 5. South Australian Health and Medical Research Institute (SAHMRI), Adelaide. 6. Hillside Campus, Cold Spring Harbor, NY. 7. Blizard Building, London. Drawings by the research team.

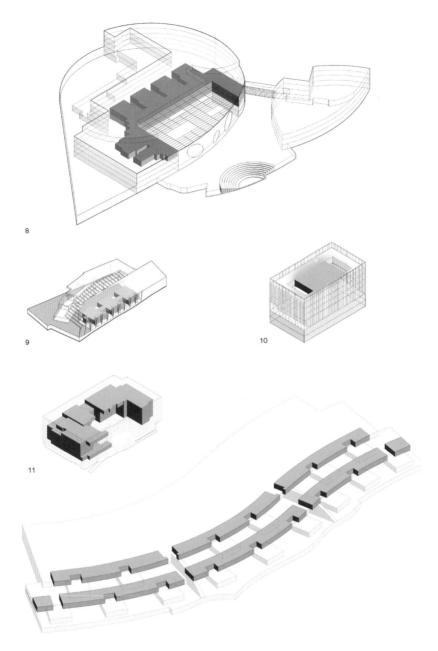

Figure 1.7b
8. Centre for the Unknown, Lisbon. 9. J. Craig Venter Institute (JCVI), La Jolla, CA. 10. Fabrik-straße 22, Basel. 11. Translational Research Institute (TRI), Brisbane. 12. Janelia Research Campus, Ashburn, VA. Drawings by the research team.

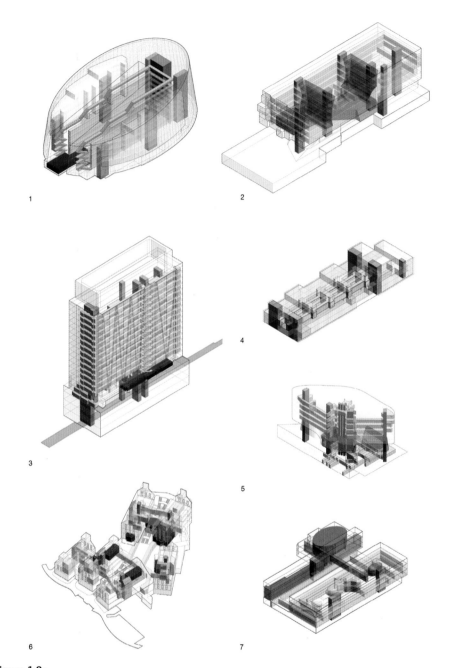

Figure 1.8a
The organization of vertical and horizontal circulation in laboratory buildings also exhibits great inconsistency. 1. PRBB, Barcelona. 2. PSB, Vancouver. 3. Belfer Building, New York. 4. Navarra BioMed, Pamplona. 5. SAHMRI, Adelaide. 6. Hillside Campus, Cold Spring Harbor, NY. 7. Blizard Building, London. Drawings by the research team.

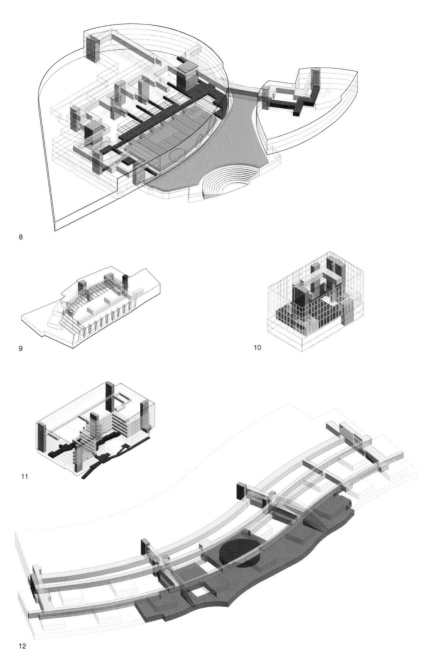

Figure 1.8b
8. Centre for the Unknown, Lisbon. 9. JCVI, La Jolla, CA. 10. Fabrikstraße 22, Basel. 11. TRI, Brisbane. 12. Janelia Research Campus, Ashburn, VA. Drawings by the research team.

proliferation of understandings and explorations. Such an investment focuses less on the "meaning" that is constructed in the relations of science and architecture, than on the operation of the construction itself: How does it perform? What does it do? Under what circumstance is it operative? How does the architecture of the laboratory construct the science that takes place within it?

Book Structure: Lines of Inquiry

The following chapter, "Oratory," details our inclusive approach to evidence. It introduces the importance of rhetoric and gossip in the construction of science, a discipline purportedly concerned only with facts. Through the Salk Institute for Biological Studies (1965) we introduce the history and tensions between knowledge and subject formation. It is fitting that we begin with the Salk Institute for it is a watershed in the design of laboratories and has been central to laboratory studies. It was here, during the late 1970s, that Latour first observed scientists as if they were an exotic tribe embarking on tasks of knowledge creation.

Following these two introductory chapters, our discussion is based on selected examples and themes. Although projects have been chosen to elucidate a single line of inquiry, the themes and ideas around each relate to other laboratories. In discussing these connections, we contextualize our position and acknowledge the relevance of other examples not within the scope of this book. The projects we discuss are best thought of as singular specimens standing in for a species.

Chapters 3, 4, and 5 are concerned with "Boundaries":

Chapter 3, "Islands": Through the figure of the island we establish the significance and effects of practices of isolation and insulation in the life sciences. We use philosopher Peter Sloterdijk's historical and epistemological account of practices and spaces of isolation to situate the development of the laboratory. We take Janelia Research Campus (2006) in Ashburn, Virginia, by Rafael Vinoly as a case study of the geographic, architectural, and rhetorical value of isolation and insulation.

Chapter 4, "Transparency": With its curved and crystalline envelope, the South Australian Health and Medical Research Institute (SAHMRI, 2013) in Adelaide, Australia, resembles a giant, prickly, glass caterpillar. It was designed by Woods Bagot for a difficult site, which straddles multiple railway lines across from the city's tattoo and massage parlors. Funded by the state, and with links to a public hospital, it exemplifies the use of glazing as the material equivalent of institutional accountability. The SAHMRI dramatizes the act of seeing-through, making optical penetration a game of legitimization and persuasion. This chapter also notes the architectural shifts evident in the proposed second-stage, the SAHMRI2 (2021).

Chapter 5, "Unbounded": The David Chipperfield-designed laboratory Fabrikstraße 22 (2010) on the campus of the pharmaceutical giant Novartis in Basel, Switzerland, is quiet on metaphors and formally conservative. The innovation is inside. In this laboratory, Novartis has taken the radical step of doing away with the physical separation of the wet laboratory in favor of a risky engineering solution that relies on exhausting biological contaminants. We argue that it is a prescient building, demonstrating the shift towards a dispersed and virtual mode of experimentation.

Chapters 6, 7, 8 and 9 are focused on the "Expression" of scientific content:

Chapter 6, "Enunciation": The inclusion of architecturally scaled images and objects, extrapolated from science, is nowhere more insistent and joyously celebrated than at the Blizard Building (2005). Designed by Will Alsop and located in a gritty inner-city area of London, it consists of a double-height rectangular prism through which can be seen brightly colored pods suspended over a subterranean laboratory. Around this building we structure a discussion about the compensatory and rhetorical nature of expression and analogy in laboratory architecture.

Chapter 7, "Excess": Few biosciences laboratories admit to their reliance on animal subjects in experimentation. The architecture of the NavarraBioMed (2011) a laboratory in Pamplona, Spain, is said by its architects, Vaillo and Irigaray, to have the skin of a polar bear and resemble a camel in silhouette. Through it, and the writings of Georges Bataille, we explore concepts and practices of excess, ornamentation, and sacrifice in the perplexing economies of the biosciences.

Chapter 8, "Deep Time": The Pharmaceutical Sciences Building (2012) at the University of British Columbia in Vancouver, Canada, is claimed by its architects, Saucier and Perrotte, to be based on trees with entangled roots and branches. Like a fossil, the structure defines the negative space of a tree within a rock-like form. Through this building, and the writings of Roger Caillois and Bruno Latour, we interrogate organic metaphors and, more importantly, the aesthetic and categorical distinctions that are made between nature and culture, plants and minerals, animals and humans.

Chapter 9, "Floating": The Centre for the Unknown (2010) on the waterfront of Pedrouços, Lisbon, in Portugal, is one of the last projects of the late Charles Correa. A composition of abstract, curvaceous forms and hard-landscaped open spaces, the Centre for the Unknown exhausts architectural representation, a situation underscored by its name. With its paved plaza, outdoor amphitheater, waterfront restaurant, and terrace bar serving mojitos, this translational research complex resembles a seaside resort. Through it we postulate the effects of Roland Barthes's concept of the floating or empty signifier.

The last three chapters consider "Socialization," which concerns the construction of the scientist and the expression of sociality:

Chapter 10, "Symbiosis": The Hillside Research Campus (2009) at the Cold Spring Harbor Laboratory (CSHL), Long Island, New York, consists of a group of six, colorfully painted, gabled buildings. The campus mimics the architectural style of the picturesque fishing village that once occupied this place. Designed by Centerbrook Architects, the Hillside Research Campus and other recent additions look to the casual eye to have been built decades earlier. The nostalgic architecture of the CSHL supports a symbiotic relationship between its scientific community and its local philanthropists. The CSHL's architecture, governance, and funding arrangements open up questions about the ways in which science absorbs and justifies the excesses of capitalism, as well as the personal impact of its long-serving director, James Watson.

Chapter 11, "Aggrandizement": On the West Coast of the United States is one of the world's first carbon-neutral laboratories, designed by Zimmer Gunsul Frasca. Like a palace, the J. Craig Venter Institute (JCVI, 2005) in La Jolla, California, is an exercise in celebrating its namesake. The chapter raises questions about the intended audiences and messages of the JCVI's architecture and interior design. We consider the location and furnishing of Craig Venter's study, as well as popular media representations of the scientist as scholar in situ.

Chapter 12, "Investment": Two speculative laboratory real estate developments in New York City—Harlem Biospace and the Alexandria Center—allow us to explore the ways in which design is used to target distinct scientific market sectors. Harlem Biospace (2013) is a tiny "co-working" laboratory for rent in a renovated brick 1930s warehouse. BAM Architecture Studio is the architect of record. The Alexandria Center for Life Science, designed by Hillier/RMJM, is a large development of three towers. Two of these were completed in 2014 and are now leased, as intended, to large pharmaceutical companies and their R&D departments. Whereas the Harlem Biospace was the home of the hipster entrepreneur Nina Tandon of the biotech start-up Epibone, the Alexandria Center caters to the anonymous employees of established corporations. Both complexes represent a privatized research landscape and the architecture that at once accommodates and constructs the contemporary scientist.

The laboratory is a particularly complex and contemporary cathedral. It is a place where biological materials, facts, and artifacts circulate and occasionally take form. The laboratory is a construction of carefully articulated boundaries and borders, layers and envelopes that at once secure the reproducibility of scientific experiments and minimize the risk of unwanted incursions or escapes. In doing so it comes to constitute a world of its own with its own esoteric system of protocols, policies, and procedures that tend to be what we think of as the laboratory's *inside*. But the laboratory is also a place that speaks loudly to its outside. The architecture of the biosciences helps give external expression to science. It sometimes does this by producing images that are identifiable and analogies that serve educational purposes writ large on walls and in volumes. The architecture makes visible the minutiae of that which largely occurs beyond the realm of human sight. These images help secure science's place in both remote and urban contexts, in public perceptions and imaginings, and in financial and philanthropic markets.

This book thus explores a rather grand experiment. It is an experiment that negotiates both scientific logic and faith. The *act* of committing vast resources into the buildings of twenty-first-century biosciences laboratories may indeed be a declaration of faith in science and its potential discoveries, but it is also a declaration of faith in architecture's power to accelerate science.

2   Oratory in the Lab: Gossip as Evidence

The word *laboratory* derives from the medieval Latin *laboratorium*, "a place for labor or work," from the Latin *laboratus*, the past participle of *laborare*, "to work." To labor is to toil with the body. The word *laboratory* first came into use in the seventeenth century and referred to a room or a building set apart for scientific experiments. Here, there was a component of physical toil in the lifting of cauldrons, the stoking of hot furnaces, the forging and forcing of heavy equipment. *Oratory*, in its early use in the fourteenth century, referred to a small chapel, or a place for prayer. Since the late sixteenth century, *oratory* has been associated with the art of speaking eloquently in public. It is derived from the Latin *oratorius*, "of speaking or pleading, pertaining to an orator," and from *orare*, "to speak, pray, plead." It is accidental, and of no etymological consequence, that the word laboratory can be neatly dissected into *lab* and *oratory*.[1]

Yet, it is significant that there is a division between the labor of the laboratory and the speaking, pleading, persuading, exhorting, explaining, arguing, negotiating, defending, gossiping, sniping, flattering, chattering, interjecting, complaining, and proclaiming undertaken by the scientist. The contemporary laboratory building separates technical activities from intellectual contemplation, the administration of the experiment from its theorization. It is a division repeated in the professional distinction between the technicians who work with instruments and matter and the lead scientists and research directors who work with words and data. The latter command higher salaries and are in a stronger position to shape the direction of scientific research. The division is mirrored, too, in architectural labor and investment. The planning of the laboratory proper is assigned to specialized firms, while architectural practices of greater renown apply their signature forms to the surfaces and spaces of socialization and public exposure.

The rhetoric around biosciences laboratories today is that they are places designed to allow "conversations to happen that wouldn't otherwise take place in a thousand years."[2] Such statements are typical in the discourse surrounding the new generation laboratory buildings. The design of the technical spaces and services go unremarked—despite

the fact they are critical to best-practice experimentation and costly in their imminent obsolescence. Instead, as we suggested in the introduction, the research building
that enfolds the wet laboratory is aimed more toward supporting managers in their
new objective, that of eliciting creativity and conversation. While technicians are not
explicitly excluded from this objective, the alleged relationship between impromptu
chats and the exchange of ideas leading to discovery may suggest otherwise. To speak,
to write, to argue are the domain of the scientist as a knowledge worker, not a laborer.

The "rhetoric of science" is, as Jean-François Lyotard discerned in *The Postmodern Condition: A Report on Knowledge* (1979), inseparable from power and wealth. Indeed, Lyotard
proposed that at the end of the millennium "the games of scientific language become the
games of the rich, in which whoever is wealthiest has the best chance of being right."[3]
The wealthiest nations and corporations have the ability to marshal the greatest resources,
just as eminent and charismatic scientists can extend their influence within the domain
of the organization and the landscape of funding. The positive reception and impact of
discoveries, their "rightness," requires the weight of resources and reputations. Individual
standing bears on the capacity to socialize one's findings and secure the trust and respect
of colleagues. In this regard, the published findings and the discussions had over lunch
are each part of the games of scientific language. Indeed, architects and their clients are
banking on the seamlessness between informal conversation and more formal discoveries.

While dining in the café-styled, subsidized canteen of the Max Planck Institute of
Molecular Cell Biology and Genetics in Dresden, we overheard a heated discussion
between scientists about the marital trials of celebrity chef Nigella Lawson and businessman Charles Saatchi. At the time, the couple were a fixation in both tabloid and
broadsheet newspapers. While the subject has no bearing on scientific research, such
casual conversations are about more than the exchange of content. Scientists become
members of workplace communities through friendships and everyday connections
and, from there, contribute to the larger construction of science as a profession. Yet science, as an epistemological form, *depends* on the elimination of the social dimension
of language. Its claims to truth rest on the exclusion of gossip, social relationships, the
inequities of funding, and access to prestigious institutions. This is where the story of
architecture and oratory starts to get interesting. It begins with the Salk Institute for
Biological Studies (1965) in La Jolla, California.

Latour in the Laboratory

As a young academic in his twenties, Bruno Latour spent between October 1975 and
August 1977 at the Salk Institute, the research center founded by Jonas Salk. Salk was the
virologist who had presented the first effective vaccine against paralytic poliomyelitis in

1955. Architects have come to know of the Salk Institute not because of its founder, or the research that has been done there, but because of its buildings. These were designed by Louis I. Kahn and considered a masterpiece of twentieth-century design. Construction was completed in 1965, after six years of successive design options being debated between the architect and client—a process prolonged by Kahn's constant emendation.

On a spectacular site in La Jolla, it consists of two six-story laboratory wings that form the north and south boundaries of the complex. The alternation of levels containing laboratories with infrastructural floors was a key innovation that allows for unobstructed layouts and ease of building maintenance. Kahn's achievements, however, are not merely pragmatic. He was inspired by the monastery of Assisi, which he had visited in 1954. Each laboratory wing shelters an inner row of semidetached towers of individual cell-like studies (figure 2.1). The studies, connected by bridges to the laboratories, have oblique views to the Pacific and face each other across a monumental travertine courtyard. The courtyard is bisected by a long and narrow rill of water that gives the impression that it flows into the Pacific Ocean below. The buildings are made of in-situ concrete, marked by the grain of the plywood forms used in construction, and accented

Figure 2.1
The cell-like offices for the scientists facing across the plaza of the Salk Institute of Biological Studies (1965) in La Jolla, California. Architect: Louis Kahn. Photograph by Doug Letterman.

by window frames and screens of unpainted teak. Sunk below ground level and hidden behind the scientists' studies, the laboratories are not visible from the courtyard. The invisibility of the laboratory function, along with the concealment of services and the textual, almost handmade, aesthetic of raw materials, departs radically from the contemporaneous expression of laboratory buildings as technical objects.

The Salk Institute was to be a place of spiritual contemplation for an elite brotherhood, more monastery than cathedral. Nevertheless, it was open to people with diverse beliefs and ideas. Long before "interdisciplinarity" became a fashionable idea in academe, the Salk Institute brought people from outside of science to work alongside its researchers. Novelist Michael Crichton, linguist Roman Jakobson, and sociologist Edgar Morin each spent long periods at the Salk Institute in its first decade. The novelty of Latour's relationship to the scientists was that he made them the subject of his studies. He planned to observe scientific work as if it were the rituals of an unknown tribe. In fact, while the laboratory and its activities were unfamiliar to Latour, the scientist who had invited Latour to conduct *an epistemological study* of his new laboratory at the Salk Institute was personally known to him. The physiologist Roger Guillemin and Latour both came from Burgundy, France.[4] Guillemin's interest in the history of scientific research had prompted the invitation. Latour was far from alone in the jungle with this tribe—there was already a precedent for European thinkers visiting the University of California at San Diego (USCD) in the 1970s. Jean-François Lyotard, Jean Baudrillard, and semiologist Paolo Fabbri each attended social gatherings at the Salk Institute.

With co-author Steve Woolgar, Latour reported his findings in *Laboratory Life: The Construction of Scientific Facts* (1979). In their report, Latour is referred to in the third-person as "the observer," as if he were a *fictional character*.[5] (Lanky and patrician, Latour is hard to imagine as a silent, unnoticed witness.) Latour and Woolgar's first observation is that scientists are "compulsive and almost manic writers" and that "a vast body of literature emanates from within the laboratory."[6] The laboratory is "a hive of writing activity"[7] and "a system of literary inscription."[8] The focus on writing is, they admit, a crutch. As nonscientists, they were unable to recognize and comprehend the puzzling mixing of liquids and gases, the strange images, the unknown vocabulary, but "by pursuing the notion of literary inscription, our observer has been able to pick his way through the labyrinth."[9] That is, Latour and Woolgar rendered the laboratory in their "own terms."[10]

Latour and Woolgar propose that science rests on texts, diagrams, and other modes of representation, which leave considerable room for uncertainty and contestation. They conclude that "scientific activity is not 'about nature,' it is a fierce fight to *construct* reality. The *laboratory* is the workplace and the set of productive forces, which makes construction possible."[11] Science is interesting for Latour not because it is outside of society, but because it isn't. Latour and Woolgar's claim that science is founded on storytelling was protested by many from within science who found it an absurd attack against the repeatability, transferability and solidity of the experimental method.[12] After Latour published *Science in Action: How to Follow Scientists and Engineers through Society* (1987), the protests turned ugly. Paul Gross and Norman Levitt, the authors of *Higher Superstition: The Academic Left and Its Quarrels with Science* (1994), described Latour as "a Panurgian imp, come to catch all those solemn scientists with their pants down, a project that delights his largely antiscientific audience."[13] The rebuke is hysterical. A prurient reader looking for an exposé of the private parts of science will be bitterly disappointed. Latour and Woolgar aspire to observing only the routine and normal practices of science at the bench, avoiding the temptation of "instances of gossip and scandal" and "sociological muckraking."[14] This aversion is repeated several times. Latour and Woolgar suggest that scientists might be suspicious of nonscientist investigators—historians, philosophers, sociologists of science—who are assumed to be fixated on "the seedier aspects of scientific life."[15] A scientist might then be prone to offering "the fodder deemed most appropriate for such investigators" in the form of "tales of scandal and intrigue."[16] That is, Latour and Woolgar blame the ubiquity of stories of treachery and deception in the history of science on scientists and their mistaken belief that this is what laypeople want to hear. In a footnote, Latour and Woolgar complain that in interviewing the scientists at the Salk Institute, a major problem was "the pressure exerted by informants on the observer to acquire the information they think he wants to hear. This is why we heard so many stories about the politics of the laboratory and why we decided not to use them."[17] So much for the neutral observer.

Latour confided to Fabbri, early into his time at the Salk Institute, his overwhelming impression of the scientists was that "they fight all the time."[18] However, this antagonistic view is largely suppressed in *Laboratory Life*, and much of what the sociologists heard was not repeated.[19] Latour and Fabbri published a provisional piece of research, titled "The Rhetoric of Science: Authority and Duty in an Article from the Exact Sciences."[20] It brought Fabbri's close semiotic reading of a scientific text together with the sociological approach that Latour was then engaged in at the Salk. The two pointed to the layers of texts that serve as mutual referents for any scientific text, and the "economy of credit" that establishes the professional and intellectual authority of the scientist. Unusually, the essay was published with a postscript summarizing Guillemin's objections. Guillemin protested that scientists do not create, but rather discover and *build things*. He condemned Latour and Fabbri for attributing everything to "personal motives, as though scientists are shifty and cunning."[21] He further asserted, "We do research, we're not playing around."[22] It is possible Guillemin's protestations led Latour to exclude from the book the personal conversations and negotiations he heard in the laboratory. Jonas Salk applauded the exclusion. In his preface, Salk wrote: "The book is free from the kind of gossip, innuendo, and embarrassing stories, and of the psychologizing often seen in other studies or commentaries. In this book the authors demonstrate what they call the social construction of science by the use of honest and valid examples of laboratory science."[23]

The aversion to gossip deserves interrogation. Gossip requires the speaker to have beliefs about the truths of their claims, thus distinguishing it from slander. It is characterized by its furtive quality, for the speaker knows that the subject of the gossip would not like what is being said. Gossip is motivated by the intention to produce pleasure in the speaker and his or her audience. To engage in gossip is pleasurable, and (perhaps because of this) it has been considered morally suspect. Gossip, along with testimony, opinion, fiction, memoirs, diaries, and oral histories, are genres of knowing that sit outside traditional epistemology with its emphasis on public, neutral standards of knowledge. At least this is the case until the 1970s, when the truth claims of texts and assumed stabilities of language were convincingly challenged by poststructural theorists.

In gossip we are confronted, Irit Rogoff enthuses, with a form of knowledge that "turns the tables on conventions of both 'history' and 'truth' by externalizing and making overt its relations to subjectivity, voyeuristic pleasure and the communicative circularity of story-telling."[24] For Rogoff, gossip offers an opportunity to decouple accounts from the authority of empirically sanctified experts and produce counternarratives, which expose the gendered writing of history. Gossip "negates the scholarly distanciation between what is said, who it is said by, and who is being addressed."[25] Gossip operates much like opinion. For the philosopher Gilles Deleuze and the psychotherapist Félix Guattari, "Opinion is the rule of correspondence of one to the other; it is a function or a proposition whose arguments are perceptions and affections, and in this sense it is a function of the lived."[26] Opinion, much like gossip, may enliven, but it has its risks, "either leading us back to the opinion from which we wanted to escape or precipitating us into the chaos that we wanted to confront."[27]

The importance of gossip for science could be argued in terms of the constitution of its disciplinary boundaries, through the process of negative differentiation, as a superior discourse. Scientific facts exist *because* they eschew gossip and the moral failures it may signal. Whereas gossip cannot be tracked back to an author and cannot be verified, scientific knowledge poses itself as the opposite. While this characterization of science is not nearly as elementary as Friedrich Nietzsche's discussion of a "gay science," nor as abrasive as Paul Feyerabend's reference to science as a "particular superstition," it has its critics.[28] Architectural history and criticism have, likewise, largely repudiated gossip and other minor texts, personal accounts, and speech events so as to claim the status and respectability of a scientific discourse. Excluded, too, are the topics that occupy gossips: who took advantage of whom; whose dirty money paid for what; what lies were told and motives concealed. Suzannah Lessard's 1997 memoir of her great grand-father, the architect Stanford White, who was murdered by the husband of one of his lovers, is a brilliant account of the continuity between architectural and sexual fixa-tions, and of a kind of writing that is both personal and historical. *The Architect of Desire* was a bestseller, but the book and its themes have remained well outside recognized architectural historiography.[29]

The Salk Institute's Architectural Absence

Curiously, architecture is accorded a similar status to gossip in accounts of its role in the production of scientific knowledge. For Latour and Woolgar, we note the labora-tory is defined as a "set of productive forces" that make the *constructions* of science pos-sible. But they stop short of suggesting the laboratory itself *constructs*. On architectural descriptions or qualities, Latour and Woolgar are silent. It is not that they are blind to Kahn's design. In his notes taken during the field research, which were published in the French translation of *Laboratory Life* (1986), Latour describes the building as a mixture of "a Greek temple and a mausoleum."[30] However, in assembling their argument about science and its social construction, he and Woolgar fail to speak of the architecture's material and spatial characteristics, just as they fail to speak of gossip. Instead, they acknowledge and argue for the exclusion of gossip, allowing the reader to recognize its absence. The discussions between scientists in *Laboratory Life* are noted as having taken place in an office, a laboratory, or a lobby, but these remain flat denotations linked to a crude plan diagram. Overlooked are the ways in which the hierarchical organization of the Salk Institute, its monumental character and overt symbolism in the landscape, frame its research and researchers. Its location in San Diego, among other research institutions and wealthy neighbors, is also deemed irrelevant. Architecture is not recog-nized here as one of the attributes or artifacts lending its scientists the symbolic capital and authority that Latour and Fabbri call "scientific credit."[31] Architecture is relegated to the same realm as gossip—circumstantial and irrelevant to the construction of texts from which the "truth-effect" of scientific knowledge arises.

Ignored, too, are the ways in which architectural spaces sponsor different conversations, at the same time as these conversations are demarcated as "inside" the dominant discourses or excluded from them. It is telling that many of the words used for gossip denote a location. "Dirty laundry," hearing rumors through "the grapevine" or "the clothesline," "back-fence talk"—all evoke the places in which people have gathered to exchange stories. Even "muckraking" takes place in a specific place. Today's architects of workplaces, as we will return to, have developed an entire vocabulary for sites and architectural techniques intended to provoke spontaneous informal conversations in pursuit of what, in 2009, MIT Media Lab's Alex Pentland coined the "water-cooler effect." Gossip around the office water-cooler is the twentieth-century equivalent of our favorite term for gossip: "scuttlebutt," the nautical slang for gossiping around the barrel of fresh water on a ship.

The apparent lack of interest in the Salk Institute's aesthetic presence that Latour and Woolgar seem to display is incomprehensible for architects, for whom it is primarily a site of pilgrimage. To see it in the flesh is a rite of passage. The building's hidden laboratories simply do not matter, and the pilgrims neither know, nor care about, the research being done there. Indeed, if you look for images on the Internet using the search terms "Salk Institute" the first thirty-five photographs are of the plaza, and people are absent in every image. In the plaza, we have seen architects jostling each other for position, armed with cameras boasting lenses as long as their forearms. Crouched low, with legs astride the channel of water, each of these devotees wants to capture an image like that photographed by Ezra Stoller in 1977 (figure 2.3). The competition is most ferocious on the day of a vernal equinox, when the setting sun aligns with the water (figure 2.2). Stoller was likely there at the same time as Latour, but neither notes the presence of the other; each was focused on their object of fixation. The situation is echoed today in architects' and scientists' mutual lack of interest in each other, which plays out each day at the Salk Institute.

Modeled on the Alhambra's fourteenth-century Patio of the Lions (Patio de los Leones), Granada, the plaza was designed in collaboration with the Mexican architect Luis Barragán. The unfurnished plaza frames expansive views of the Pacific Ocean and, in Barragán's words, makes of it a "façade to the sky."[32] The architecture critic Herbert Muschamp called it "the most sublime landscape ever created by an American architect."[33] Robert Venturi and Denise Scott Brown go further in their patriotism, claiming that it is "poignantly American as it frames the sea and the land where the old western frontier ends and the new eastern frontier begins."[34] Critic Paul Goldberger, writing for the *New York Times*, gushes that the plaza "is both the symbolic center of the village that the Salk Institute is intended to be and an expression of man-made, urban community

Figure 2.2
Photographers at the plaza of the Salk Institute for Biological Studies (1965), on the Northward equinox, March 20, 2015. Architect: Louis Kahn. Photograph by Steve Aldana.

set in magnificent juxtaposition to the infinite openness of the ocean. It is surely the greatest outdoor room in American architecture since Thomas Jefferson's Lawn."[35] The architect Tod Williams supposes it a "gracious place of discourse, reflection, and discovery."[36] Gossip, it seems, would not comfortably occupy such a refined space.

Kahn's reinterpretation of a European village and monastery, his use of tactile materials and details, and his attention to the topography of the bluff and the coastal environment heralded a reconciliation with architectural history after modernism's machine aesthetic. Then, and now, the Salk Institute is considered by architects to be an exemplary realization of the principles of *place-making*. It has always been assumed that the scientists there are ennobled by their habitat. The assumption is wrong. Latour's portrayal of life at the Salk Institute, even without the gossip, revealed it as "a savage brawl in which, from day to day, the dominant chieftain is he who assembles, by dint of wealth, prestige, and warrior cunning, the biggest and nastiest gang of henchmen."[37] His revelations about the Salk Institute's internally competitive culture did not shock the scientists who worked there. Nobel laureate Paul Berg told Ann Gibbons that the Salk Institute "has become anything but that early vision of a think tank or an Institute for Advanced Studies. They're scratching to survive just as much as the rest of us."[38]

To explain how the architecture of the Salk Institute became invisible to the historians and sociologists of science—even as they worked to reveal the myths of science's neutrality and truth-telling—is a long story. Part of the story relates to the raison d'être of the laboratory, which is, of course, to exclude everything that is not a part of the experimental set-up—vibration, heat, humidity, light, dust, noise, bacteria, gossip and opinion, and architectural tourists with their telephoto lenses. The idealized neutral space extended and paralleled the scientific project of arriving objectively at verifiable facts. Scientific facts were to be unimpeded by prejudice and hearsay, uncolored by interpretation and intuition, and untouched by the politics of gender, class, age, and race, which, outside of the laboratory, give authority to subjects making competing truth claims. The foundational mythology of the laboratory, Thomas Gieryn suggests, might make "science-buildings into a 'hardest case' for demonstrating that space and place are fundamentally involved in the reproduction of social life."[39] It's a challenge that has defeated even the most determined historian of science.

The Placelessness of the Laboratory

In late 2007, Professor Emeritus Robert Kohler sat in his over-heated office at the University of Pennsylvania, trying to concentrate on Emily Pawley's draft of her thesis on New York farms in the mid-1800s.[40] His thirteenth dissertation student, Emily may be the last. After thirty-two years in Penn's Department of History and Sociology of Science, Kohler is recently retired. Yet for the past two years he has continued to come to the office each day. Scrawling in the dissertation margins in red pen, he welcomes the interruption of an email. Looking back on this moment, Kohler writes: "When an email message from *Isis* editor Bernie Lightman appeared in my inbox, asking if I would take part in a Focus section on laboratory history, I was surprised. Not because the project seemed to me odd or unnecessary—I had been feeling for some time that lab history was being sidelined—but because I had no inkling that others felt the same way."[41]

Isis is the official journal of the History of Science Society. Established in 1912 (long before the other ISIS, the Islamic State of Iraq and the Levant and before the History of Science Society itself), it is also the oldest journal in its field. On the eve of his retirement in 2004, Kohler was awarded the society's Sarton Medal for exceptional scholarship over a lifetime, yet, still, he yearns to be part of a larger conversation. Lightman has asked Kohler to write an introductory essay on the laboratory, a kind of summing up of the field. We imagine the exercise feels to Kohler oddly like writing a memorial valediction, rather than a return from the sidelines. He ends up writing a short historiography on the laboratory that begins with his own career move from chemistry in 1970.[42] Soon after that move, and following the appearance of Latour and Woolgar's study, among other texts, the laboratory seemed poised to become the epicenter of investigations in the history and sociology of science. Twenty years later, in 1995, Karin Knorr Cetina was still optimistically proclaiming the emerging field of "laboratory studies." Yet another decade on, Kohler laments that there is still no comprehensive history of the laboratory. What is missing he writes, "are studies of labs as social and cultural infrastructure.... The lab has reverted to what it was in the bad old days: a neutral, invisible stage for scientific fact making."[43]

Kohler proposes that the *placelessness* of the laboratory is confirmed by its dispersal, its ubiquity, and its cultural plasticity. He selects a single illustration to make his point—the iconic photograph taken by Ezra Stoller of the Salk Institute buildings, from the axial center of the famous plaza looking out to the horizon of sea and sky. The photograph, published courtesy of the Salk Institute (who clearly haven't seen what is coming), is reproduced twice: first on the cover of *Isis*, and then in Kohler's essay, where it is captioned: "Laboratory high modern ... Louis Kahn's celebrated design epitomizes the ideal of 'placelessness' that characterizes the modern lab."[44]

There are several unflattering observations that one could make about Kahn's Salk Institute, but "placelessness" will come as a surprise to architectural readers. Yes, Kahn's design has been difficult to adapt to the needs of its scientists. The external corridors, alongside the interstitial service floors, are jammed with the specialized refrigerators and equipment used by researchers in the biosciences—equipment that in new research buildings is securely housed alongside bench space (figure 2.4). It is also true that it has been immeasurably difficult to expand given that its symmetrical organization and basement laboratories defy easy addition. Its designation as a historical landmark in 1991 hasn't helped. In 2006, the entire twenty-seven-acre (eleven-hectare) site was deemed eligible by the California Historical Resources Commission for listing on the National Register of Historic Places. Demonstrating sensitivity to the architectural significance of the old, new buildings completed in 1996 by David Rinehart and John MacAllister of Anshen and Allen are sited at a polite distance and with a considerable portion below ground. Even so, they were poorly received by architectural critics who found them "obsequious," "sycophantic," and bland.[45] The Salk Institute's status as a destination for architects and tourists has also proved something of a double-edged sword. That this form of tourism is simultaneously a generator of revenue and an intrusion is perfectly summed-up by a sign fixed by a scientist to the window of his laboratory. It reads: "Please FEED THE SCIENTISTS."

From the outset, the plaza at the heart of Kahn's design for the Salk Institute proved more photogenic than socially enriching. Even Thomas Leslie, an ardent admirer of

Figure 2.3
The cover of the *Isis* special issue on laboratories, vol. 99, no.4 (2008), featuring Ezra Stoller's 1977 photograph of the Salk Institute for Biological Studies.

Figure 2.4
External corridor of one of the interstitial service floors lined with equipment at the Salk Institute.
Photograph by authors (2013).

Kahn's ability to meld technical and aesthetic demands, admits that "from the beginning, almost no one sat in the courtyard."[46] Through the removal of plants from its center (as Barragan had originally envisaged), biological life is also cast out, only to reappear in molecular fragments on the laboratory bench. The sunken courtyards, which flank the peripheral and subterranean laboratories, are very much lived in—with surfboards and wetsuits set out to dry and groups of scientists relaxing around plastic outdoor tables (the sort no architect would specify), but this life remains out of view (figure 2.5). The terrace, upon which the scientists gather to lunch in the Californian sun, also sits out of sight of the touring architects on a level below the plaza. Overhead, and also out of the camera's frame, military jets make raucous sorties from a nearby air force base. The silence of the photographs of the plaza is a seductive fiction. So, too, is the idea of the plaza as a great public space, for this is a private institution that can be reached only by car, where access is carefully limited and monitored.

Nevertheless, it is difficult to reconcile Kohler's view of the Salk Institute's "placelessness" with the ways in which that term is used in architecture to refer to the generic

Figure 2.5
Sunken laboratory courtyards at the Salk Institute for Biological Studies (1965), La Jolla, California. Architect: Louis I. Kahn. Photograph by authors.

and repetitive indistinguishability of suburban shopping centers and airports. The peri-urban landscape of big box retail and business parks that lines Interstate 5 (I5) between San Diego and Los Angeles is consistent with what architects and urban theorists call "non-place" and "terrain-vague" and where they discern a dispiriting placelessness. The Salk Institute—once one leaves its surrounding car parks behind—is a reassuring oasis in this environment. Paradoxically, it depends on the I5, a *catalyst* for the development of one of North America's largest biotechnology enclaves.[47] Nevertheless, in the parallel universe that is architecture, Kohler's accusation is tantamount to heresy. It is perhaps fortunate for him that very few architects read *Isis*; if they did, the emails in his inbox would have been less than welcome. For Kohler, placelessness is not necessarily pejorative, nor is it essentially architectural. Rather, it is a quality that allows modern science to flourish. He writes:

> Placelessness marks lab-made facts as true not just to their local makers but to everyone, anywhere. It marks the lab as a social form that travels and is easy to adopt, because it seems rooted in no particular cultural soil, but, rather, in a universal modernity. Placeless means

dispersible.... Labs are of course not literally "no place." Placelessness is a social or symbolic attribute: an artfully constructed yet highly potent and useful social fiction.[48]

This is not a new idea. In fact, it was the sociologist Émile Durkheim who first distinguished religion from science in terms of the localization of the former and the placelessness of the latter: "Religious beliefs in the less developed societies show the imprint of the soil upon which they are formed; today, the truths of science are independent of any local context."[49]

What is fresh in Kohler's formulation is the idea that placelessness is produced by an *artful construction* and is a quality of being able to be adopted elsewhere. Indeed, one can read Kohler's description with airports in mind and note the value of "a social form that travels and is easy to adopt." Placelessness for Kohler is not evidence of the absence or failure of design—as architects who promote themselves as placemakers would have it; but it is a difficult argument Kohler is making. First, he proposes historians of science have failed to account for the specificity of laboratories in history and place, while at the same time arguing the laboratory is universal. Second, he sees the placelessness of the laboratory as a rhetorical fiction. This makes a little more sense in a historical context. For Kohler, nineteenth-century science is characterized by the trust placed in individual gentlemen who are members of a powerful social elite.[50] Kohler draws his argument from fellow historians of science, such as Simon Schaffer and Steven Shapin.[51] The modern laboratory, however, accrues authority through its resonance with and connections to other powerful and ubiquitous modern institutions, such as the nation-state, humanism, the middle class, and the capitalist economy. The modern laboratory, and the Salk Institute more specifically, encapsulates moral values of objectivity and individual discovery, an ethos of intellectual purity and the idea of science as a public good. It is artfully "placed" in these frameworks.

The Salk Institute's Long Shadow

In the narratives told to us by architects, scientists, and building managers, the Salk Institute has appeared regularly as a legendary but elusive character, taller in each retelling. Take for example, this story from Barcelona. On receiving commitment from the Barcelona City Council and Pompeu Fabra University for the new Barcelona Biomedical Research Park (PRBB) (2006), the Park's incoming director, Jordi Cami, and his senior management team flew to California with the architects Manuel Brullet and Albert de Pineda. Brullet, who had visited the Salk Institute decades earlier, insisted that his clients see Kahn's masterpiece first hand. What the architect was looking for in the Salk Institute—and hoping to reveal to his client—was something other than a pragmatic understanding of solutions to the laboratory program. Brullet writes:

> The influences of this building on the Biomedical Research Park are subtle and varied, yet there is one highly explicit image and that is the appearance within the "cloister" in the park of the sheet of water on the roof of the auditorium and its merger with the Mediterranean, which is indebted to the canal of water in the central courtyard of the Salk laboratories that disappears into the Pacific.[52]

The importance of the Salk Institute for architects lies in its "explicit image" and the hopeful idea that architecture is concurrent with the profound intellectual work undertaken by scientists. Brullet wanted the same visual effect of connecting the PRBB to the sky and sea. He was blessed with a waterfront site, but as it was low-lying, he created a courtyard with a small auditorium at its center on top of which he installed a shallow pool. Ultimately his efforts have been thwarted by leaks, and the client has since drained the auditorium roof of its reflective sheet of water (figure 2.6).

Likewise, the design of the J. Craig Venter Institute (JCVI), also in La Jolla, is intended to have the architectural qualities and presence of the Salk Institute.[53] Completed in 2014, it, too, is a pale imitation. Two miles south of the Salk Institute and on a smaller site, the JCVI utilizes the same raw teak walls and exposed concrete as its neighbor, but its courtyard is long and narrow, and its flanking wings are asymmetrical. A roof of photovoltaic panels obscures the sky, and Venter's large office at the prow of the building obscures the views to the sea, as if to deny all but him the pleasures of the elevated site. As will be discussed in detail in chapter 11, the JCVI is a palace for a scientist and his courtiers. It is this shift in ambition that separates it profoundly from Kahn's egalitarian monastery ideal.

For Kahn, as for Brullet and Zimmer Gunsul Frasca, the architects of the JCVI, the organization of the laboratories and offices around a courtyard aimed at a common and inspiring setting for the scientific community. Kahn wanted to capture Jonas Salk's characterization of science as akin to the monastic pursuit of higher knowledge, with distinct spaces for public ritual and private contemplation.[54] Salk believed a well-designed, or iconic, building would enhance working life and, perhaps more importantly for the longevity of the institution, attract leading scientists and express the ideals of the institution and its activities to the public. The laboratories at PRBB and the JCVI, despite their formal reference to the Salk Institute, are operating in a context where the scientific community has been transformed, as has its workplace organization. What has lingered from the Salk Institute is the *idea* that spaces can be designed to foster socialization and represent community. While Latour and Woolgar were "discovering" the social basis of science, architects such as Kahn were, at the behest of

Figure 2.6
Barcelona Biomedical Research Park (PRBB) (2006), Barcelona, Spain, looking out to sea across the now dry pool. Architects: Manuel Brullet and Albert de Pineda. Photograph by authors.

scientific managers, working in the reverse direction. Salk and Kahn were hell-bent on designing a place like a village in which "gossip" thrived. In 2014, William Brody, president of the Salk Institute, recalls that "Jonas saw the Salk Institute as an artists' colony for scientists."[55] Such ambitions have now become commonplace.

The Social Laboratory

If the Salk Institute heralded the emergence of the modern laboratory as a space of collaboration and teamwork, today these objectives have reached an exaggerated mode. Cooperation, creativity, and entrepreneurship are folded into a quest for self-realization and communication. Kohler sees this urge to communicate and form communities as one that is intimately tied to business management. By this he means not only the spreadsheets, performance measures, and soft workplace culture techniques used by scientific managers, but also the management of science through funding and policy by the state, private investors, stock markets, and philanthropists. Indeed, Kohler proposes the history of American science in the twentieth century to be inseparable from the philanthropic activities of major donors, such as the Rockefeller Foundation and the Carnegie Corporation. Private and state grant processes foster the large-scale organization of scientific research around issues deemed worthy by nonscientists.[56] This is what Kohler sees in the Salk Institute—its disciplining of the scientist into organizational objectives. The dominance of short-term grants and the favoring of applied research with imminent outcomes see a disintegration of stable careers and their replacement by project-based employment patterns, veiled behind a rhetoric of autonomy and entrepreneurship. It's launched a different, less overt, modulation of the scientist as a chattering extrovert. The *labor* of the *laboratory* becomes ever more abstract.

Kohler may also have been thinking about placelessness in the context of the *exchangeability* of places in capitalism. At the time Kohler was writing his inconclusive essay on laboratories for *Isis*, it was announced that Logan Hall, the building in which he had an office, was to be renamed after the late gossip columnist and socialite Claudia Cohen.[57] Cohen's former husband—chair of Revlon cosmetics, Ronald Perelman—had donated USD 20 million to the university and acquired the right to rename the 1873 stone building. On the National Register for the University of Pennsylvania Historic District, the building had been named in 1906 in memory of James Logan, William Penn's colonial secretary and a founding trustee of the College of Philadelphia, the university's predecessor institution. Logan was a natural scientist whose primary contribution to the emerging field of botany was a treatise that described experiments on the impregnation of plant seeds, especially corn. He tutored John Bartram, the

American botanist, in Latin and introduced him to Linnaeus.[58] Logan is perhaps best known, though, for being a bibliophile, confessing once, "Books are my disease."[59] In 2008, when the name change was announced, chemistry professor Ponzy Lu told the *New York Times*, "I, as an academic, am accustomed to seeing buildings with names like Newton, Copernicus, Darwin. Then to see the name of this person … who is not associated with a pursuit of knowledge—a gossip columnist: it strikes me as being totally idiotic."[60]

The link between gossip, wealth, science, and buildings at Cohen Hall is explicit, and the story may seem a distraction to the business of architectural criticism and the operation of laboratory buildings. We tell it because it underscores the way in which buildings are political agents, avatars and unstable carriers of semiotic burdens. They are not conceived by architects operating autonomously, whose uniquely personal visions are realized, and forever preserved, in uncompromised perfection and through which they communicate meaning in a timeless and universal fashion. It may seem some of our criticisms in this book impute a failure of imagination, or resolve, on behalf of the architects of laboratories, but this is not intended. Rather, it is an effect of the tendency for the profession to overstate its agency. We insist the expression of science, through its architectural sites, cannot be reduced to notions of *artistic expression*. Laboratory buildings are material artifacts, social condensers, subjects for photography, registers of organizational relationships, containers for machines, places for gathering and gossip. Buildings temporarily solidify financial flows across the territories of capital. They are commissioned, designed, built, occupied, owned, named, and critiqued by large numbers of people, each with individual subjectivity, agendas, and social positions. Each person, be they client, critic, builder, scientist, or architect, acts through text, speech, image, data, drawing, diagram and photographic artefact—adding another layer of mediation. This is why any discussion of architecture's value, intent, and effect must be broad, intertextual, and not hastily denounce and discount the trivial. Moreover, it should look not just at the intrinsic qualities of a statement, but at its transformation by others—its uses and effects. This is why we pursue the *oratory* in the *laboratory*.

Part I: Boundaries

The raison d'être of the laboratory is enforced in the construction of boundaries: walls that insulate experiments and isolate against external materials and forces; and openings that filter the movement of everything that comes in or out of a laboratory. Such boundaries ensure the replicability of any experiment by removing those forces that are themselves not transferable. All laboratories are measured, categorized, and certified in terms of four levels of isolation. The levels correlate to the threat posed to the population by the materials within. The threat they pose may be biological, social, or imagined. Some of these dangerous materials—anthrax, for example—predate the laboratory and are brought within its realm of management and isolation. Others are made in the laboratory. Therefore, it is necessary to see the laboratory as a two-way system. Its walls are about both insulation and isolation. Insulation tends to belong to insides, and the regulation of those insides. Isolation tends to be constructed by forces of the outside and a relation to contexts. Together insulation and isolation are both the producers and products of the walls, the boundaries and membranes, of the laboratory. In parallel, Bruno Latour notes that the forces that construct the laboratory from the inside tend to be referred to as "micro." The forces that construct the laboratory from the outside are referred to as "macro," and he concludes that "no matter how divided they are on sociology of science, the macroanalysts and the microanalysts share one prejudice: *that science stops or begins at the laboratory walls.*"[1]

The construction of physical boundaries has as its correlate the desire to constantly enforce, draw, and redraw the boundaries around the disciplines of bioscience and the organizations and institutions that constitute it.[2] The following three chapters explore the complex boundaries that generate the laboratory, the disciplines of bioscience and scientists themselves.

3   Islands

The laboratory is figured, and reconfigured, in order that it might continue to generate science and the scientist. Much of the reconfiguring of the early decades of the
twenty-first century rests on a belief that the best way to accelerate discovery is to
bring researchers from multiple disciplines into a bounded space. This belief links two
key ideas: the first relates to disciplinarity, and the second relates to spatial demarcation. Disciplinarity is understood as the structural distinctions of knowledge and
information (discourse) and, following Michel Foucault, the demarcations that are
constructed by knowledge.[1] This understanding invests in the notion of disciplines as a
disciplining—a set of conditions and tendencies toward the conditioning of knowledge
and the formulation of boundaries: disciplinary boundaries. The tasks of science today
tend to be framed as responses to "big" problems, and such problems require diverse
skill sets and expertise. Co-locating researchers drawn from disciplines and organizations that have previously been isolated from each other is believed to lead to new
discoveries. Thus, pure research is brought together with clinical practice: microbiology
with epidemiology; adult cancer research with research into children's cancer; pharmacy with genetics; and mathematics with neuroscience. Scientists often refer to this
in spatial terms: the destruction of silos and the breaching of walls. Bringing disciplines
together involves the creation of new spaces in which protagonists might productively
interact. Each of these new spaces possesses an internal logic and enacts practices of
organizational affiliation and identity formation. In this way, the contemporary laboratory might be considered a set of conditions and tendencies toward the soliciting
of interactions and the formulation of boundaries: architectural boundaries. New disciplinary boundaries are constructed concurrently with geographic and architectural
boundaries. We call this the making of islands.

Janelia Research Campus

Nowhere is this practice of island-making clearer than at the Howard Hughes Medical Institute (HHMI) Janelia Research Campus (2006) in Virginia. This place was called "Janelia Research Farm" when the building we discuss in this chapter was constructed, and scientists still tend to refer to it as "Janelia Farm," so we will too.[2] Here, the planning of a new research organization, the selection of a site, and the design of its architecture were all intertwined. There were no preexisting research groups or organizations, no lobbying for space, and no current research projects to be housed. Indeed, there were no discipline groupings or researchers agreed upon at the outset. Likewise, there was no site nor any existing building. There was, however considerable economic investment. The HHMI has operated since 1953 and currently describes itself as "an independent, ever-evolving philanthropy that supports basic biomedical scientists and educators with the potential for transformative impact."[3] HHMI is the second-richest medical research foundation in the world, after the British Wellcome Trust. The HHMI has a USD 16 billion endowment, which it previously used to fund around 320 researchers at seventy universities across the United States and a hundred locations internationally.[4] The idea behind Janelia Farm was to supplement this program and enable interdisciplinary collaborations on problems that require long-term effort in excess of the term of government grants. HHMI set out to co-locate a similar number of researchers, some for periods of a few weeks, others for several years, in a well-equipped and architecturally designed facility on a physically isolated and scenic site.[5] The HHMI was in the rare position of being able to establish an entirely new research island at a capital cost of USD 500 million.

The chosen site for Janelia Farm was 281 wooded acres along the Potomac River in semirural Loudoun County, Virginia. The nearest town is Leesburg, six miles away. The property had been the country estate of Vinton and Robert Pickens, who had purchased it in 1934 while living in Switzerland to protect themselves from the devaluation of cash.[6] The property was named for the Pickens' daughters, Jane and Cornelia. The Normandy manor–style mansion they commissioned from the Boston firm of Smith and Walker in 1936 remains and is on the National Register of Historic Places. It was not a working farm—Robert was a writer and Vinton an artist—although the Pickenses grew lilies and bred horses and dogs as a pastime on the scenic grounds.[7] But Nobel-Prize winning chemist and president of the HHMI, Thomas Cech enthusiastically engaged the metaphor of a "farm," for "a farmer chooses what to plant, and then fertilizes and waters those crops to support a process of continual renewal. We will do something similar at Janelia Farm: We will identify the best scientists and nurture their research in a fertile environment."[8] Both science and the scientist were to be grown on Janelia Farm. Here, early career scientists would be quarantined from the distracting demands of teaching, performance reviews, grant writing, committee work, academic politics, and the logistics of securing child care.[9] Research funding would be internal, dependable, and generous.

The HHMI studied other examples of free-standing research communities: the Carnegie Institution of the Washington Department of Embryology, established in 1913; the Cold Spring Harbor Laboratories, established in the 1890s (and the subject of chapter 10); the European Molecular Biology Laboratory (EMBL), established in 1974; the Medical Research Council Laboratory of Molecular Biology (MRC LMB), established in 1947; and AT&T's Bell Laboratories, established in 1963.[10] While the HHMI learned from the successes and failures of these predecessors, the organization focused on the *ideal* design for interdisciplinary engagements and the boundaries of biosciences research. The 2003 *Report on Program Development* by the HHMI encapsulates the planning and progress of the Janelia Farm campus, four years after it was first mooted. Cech explains that Janelia Farm had its genesis in conversations with David Clayton and Gerry Rubin as they "thought about ways to expand the boundaries of medical research."[11] Expansion was not the only objective. Janelia Farm was also conceived as "the loosening of the boundaries between people…. Scientists have been able to cross those barriers."[12] It is not just the expansion and loosening of the boundaries of research that were considered by the HHMI. The document shows that the HHMI described in spatial and temporal terms the boundaries of both governance and architecture. Equal attention is given to both the limits of employment contracts and the divisions and connections between one laboratory and another:

> The scientific programs at Janelia Farm will be designed to further collaboration and creativity among scientists. Research teams will be kept small and team leaders will be expected to stay actively involved in bench research, not just manage it or guide it. The architectural design of the Janelia Farm buildings and its laboratories will respond to these same objectives, with both work and relaxation areas designed to promote interaction and collegiality—and discourage isolation.[13]

The choice of site for Janelia Farm campus was an interesting one, given the aim of discouraging isolation. Here geography ensures the isolation of the scientific community from the world around, while the individual scientist is compelled to take part in the community. The architecture embodies the tensions that ensue. The research buildings were designed by Rafael Viñoly after a two-stage competition (figure 3.1).

Figure 3.1
Aerial view of the Janelia Research Campus (2006), Virginia, showing the terraced and semi-buried
landscape building in the middle ground and the historic Janelia Farm mansion in the background. Architect: Rafael Viñoly Architects. Photograph by Paul Fetters, courtesy of Rafael Vinoly
Architects.

The project is described as "somewhat unique in that the planning for both the scientific program and the campus facilities has been intertwined, with each part overlap-
ping and influencing the other."[14] Viñoly understood the project was about scientific
research that might breach disciplinary boundaries. Janelia Farm would "bring together
biologists, computer scientists, engineers, physicists and chemists, free them from
many of the constraints that dominate more traditional research environments and
give them the opportunity to cultivate the new tools of biology."[15] At the same time,
they would mitigate against stagnation by introducing conferences and short-term fellowships. The community would not get stale. A fresh supply of scientists and new
ideas was guaranteed by limiting contracts for the resident scientists to no longer than
five years (there were special conditions for the initial research-leader appointments).
Janelia Farm would be a free-standing research community that grew a barn-raised
scientist.

Much of the rhetoric around the establishment of Janelia Farm centers on its novelty and its uniqueness and how it encourages interaction, but it does this by forming
a somewhat isolated community. While Janelia Farm was the first research campus for

the HHMI, it was not the first example of scientists imagining and constructing ideal laboratories as metaphoric, or real, islands. Indeed, the history of science is a populated with island-making and island-hopping; journeys from one island to another; archipelagos and reclaimed territories.

Science as an Island

The philosopher of science Michel Serres insists that science is an island of logic and rationality. This island is a pocket of order in a chaotic sea. In *Hermés I: La communication* (1969) Serres inverts the usual image we have of science as a thorough and enveloping system of logic that has built up a progressive understanding.[16] This image of science leads to the illusion that today's scientists are merely filling the small, empty gaps in that knowledge. For Serres, however, science has carved out only a small zone, a "tiny island," of understanding in what is an otherwise complex, dynamic, and irrational world. For Serres, "The 'rational' is a tiny island of reality, a rare summit, exceptional, as miraculous as the complex system that produces it, by a slow conquest of the surf's randomness along the coast."[17] Everything, including the individual, is implicated in the swirling sea of chaotic connections and complexities that constitute the world: "A macro-molecule, or any given crystallized solid, or the system of the world, or ultimately what I call 'me'—we are all in the same boat."[18] In 1991, Bruno Latour teases out the idea in a series of interviews with Serres. Latour asks about the "demarcation between what is truly science and what is truly not."[19] Serres suggests that "this border fluctuates continually, from one extremity of the heavens to the other.... Don't imagine that the sciences are like the continents surrounded by watery abysses. Not at all. They are more like oceans."[20] The influence of the idea on Latour is pronounced. Over a decade later, Latour suggests that "the world is not a solid continent of facts sprinkled by a few lakes of uncertainties, but a vast ocean of uncertainties speckled by a few islands of calibrated and stabilized forms."[21]

What we get from such an image of science is the sense of a tightly constituted form of logic, which has carved out a small and relatively stable space for itself. If science is a tiny island, then the laboratory is the system of dredging at its unstable edges and in its shallower waters, which creates a "calibrated and stabilized" boundary. The laboratory attempts to remove the contingencies and variabilities that make the laboratory's findings, in an oxymoronic manner, more universal and mobile. The laboratory enacts the deliberation over the inclusions and exclusions from science, the what and who that are truly in or truly not. (While architects, chefs, and the like frequently refer to their offices and kitchens as "labs," they are clearly not in.) The story of the relation between science and islands and islands and laboratories significantly predates Serres and Latour and is implicated in the very origins of the biomedical laboratory, both factual and fictional.

In the Same Boat

The narrative structure of H. G. Wells's *The Island of Doctor Moreau* (1896) popularized an image of biological research as isolated on islands. The main character (and supposed author of Wells's story), Edward Prendick, we are told, was "taken to Natural History" and had studied "Biology at University College."[22] An objective distance is created. A type of *scientific gaze*. Edward Prendick, found himself in the Indian Ocean, in a dinghy of the *Lady Vain*, following a collision at sea on February 1, 1887. In the introduction to Wells's novel, Prendick's nephew tells us, "My uncle passed out of human knowledge about latitude 5′ S. and longitude 105′ E., and reappeared in the same part of the ocean after a space of eleven months."[23] Prendick shared the dinghy with three others. He was also "in the same boat" as the biological science of the time. Wells knew his topic. He had studied at the University of Edinburgh, some forty years after Charles Darwin.

The dinghy of Wells's novel, Prendick's dinghy, drifted for eight days and came to rest on an island. This would have been far less ominous an event had it occurred before the return of the *HMS Beagle* and Darwin's subsequent publication in 1859 of *On the Origin of Species*. For it is Darwin who demonstrates that the island is a magnificent laboratory. Islands allow a species to be marooned from its kin and, thus, able to develop differences that come to constitute new species. These groups are removed from the exterior forces that may have taken them down another path or wiped them out altogether. They were thus subject to insulated and isolated "conditions of life," a potentially prosperous situation for a species. Though Darwin cautions about the observation, he suggests that "an oceanic island at first sight seems to have been highly favorable for the production of new species."[24] The description of these islands resonates with contemporary accounts of the biomedical laboratory—insulated and isolated and yet internally operative "for the work of improvement." Janelia Farm, likewise, frames itself in terms of novelty and differences that are entirely dependent upon the insulation of its scientists, their being quarantined from nonproductive academic labor and other organizations, just as they are isolated on a campus six miles from the nearest small town.

If the publication of *On the Origin of Species* alters the locality of the laboratory—from the basement, attic, and garden to the island—it can also be said that it extends the scientific gaze to all that inhabits islands. The slide of natural history and natural philosophy into the discipline of biology of the late nineteenth century was concurrent with the extension of science's purview—to all life, in all places, over all time. This all-consuming investment called for equally profound and all-encompassing theories. In Latour's words, "Give Me a Laboratory and I Will Raise the World."[25]

Absolute Islands: Uraniborg

The German philosopher and cultural theorist Peter Sloterdijk noted that with the advent of the Enlightenment, the project of locating, discovering, and finding literal oceanic islands became a technical-political pursuit related to the designing of islands. In modernity islands shifted "from the register of the found to the register of the made. The moderns are island-writing and island-building."[26] Sloterdijk observes the making of islands involves the simple act of isolating, and this act has become a key point of cultural and technical investment in modern times. Since Sloterdijk has identified the centrality of the process of isolating (and noting the parallel act of insulating), it is not surprising that laboratories are among his key examples of islands. For Sloterdijk raising the world via a laboratory and making islands seem to be synchronous endeavors. This idea has its origins in his short book *Im selben Boot* (*In the Same Boat*, 1993) and is extensively elaborated on across his epic *Spheres* trilogy (1998–2004).

For Sloterdijk, "Spheres are sites of inter-animal and interpersonal resonance within which the gathering of living beings engenders a plastic power. This power is such that the form of coexistence can go so far as to alter the very physiology of its co-inhabitant."[27] Every sphere is a farm, or an island of a kind. A type of physical and abstract container in which all that one perceives, and has access to, is contained. It is also a type of logical system. For Sloterdijk, the largest and most axiomatic sphere is monotheistic religion, which he calls "One Orb." It no longer holds, as Sloterdijk concedes: "The One Orb has imploded—now the foams are alive."[28] As one sphere pops, it gives birth to others. In the smaller and more particularized spheres of "foam," there is a liberation of multiplicities from the unities that otherwise contained them. In the final volume of his trilogy *Foams* (2016), Sloterdijk turns to the contemporary era to develop his theory of capsules, islands, greenhouses, and the processes of island formation.[29] It is the insulation and isolation of both islands and laboratories that operate as a common denominator. Different processes of isolating generate different types of islands. Sloterdijk identifies three specific technical types: *absolute islands*, *atmospheric islands*, and *anthropogenic islands*.

Absolute islands do not have coastlines, but instead have carefully defined, sealed, walls. Examples of the completely sealed, life-sustaining, self-regulated, and tightly controlled environment can be found in spaceships and space stations. The most important example for Sloterdijk does not, however, fly in the sky or orbit the earth, but finds itself isolated in the depths of the ocean. Sloterdijk turns to Jules Verne's *Twenty Thousand Leagues Under the Sea* (1870) to find his absolute island: "The electrically driven underwater hotel *Nautilus*, born of the great misanthrope's inventive spirit, embodies a first technically perfect projection of the idea of absolute insularity—a world model of extreme closure and introversion."[30] Captained by Nemo, an inventor, a scientist, an exile (also known as Prince Dakkar), and a leader in search of a mysterious sea monster, the *Nautilus* comes to represent all that makes an absolute island. Controlled, contained, and generating experimental outcomes that are universally applicable, the *Nautilus* constitutes its own world. The *Nautilus* is much like every wet laboratory with its physical containment levels, rules, regulations, and controls. It operates as a legal entity, a legal island—like a casino, an embassy, a refugee center, a Siberian closed city, or a space-station—with its own highly specific legal frameworks, restraints, and privileges. For Sloterdijk, this "worldwide model" serves as a diagram for the forms of insulation, isolation, and introversion to which so much of contemporary life aspires. The motto of *Nautilus* was "Mobilis in mobili" (moving amidst mobility).

To make and occupy such an "absolute island" was the aspiration of the first island-based scientist, Tycho Brahe. Brahe is best known for his 1572 discovery of a new star. It was a startling discovery because the heavens were meant to be fixed. Prior to Brahe, stars weren't born; they were eternal and part of the God-given cosmos (One Orb). His star pupil, Johannes Kepler, was to use Brahe's work in developing his revolutionary laws of planetary motion in the seventeenth century. Brahe's *Nautilus* was the mid-sixteenth-century castle Uraniborg on the island of Hvan in Denmark (figure 3.2). Brahe became the feudal lord of Hvan. Uraniborg was star-shaped in plan, with a laboratory and observatory, which were built on the island to avoid distraction. It "evoked a life of withdrawal devoted to the contemplative study and articulation of eternal verities."[31] Brahe's castle was on the island, at the center of a large walled estate, with his chemical laboratory separated from the house and located underground. Surely the highpoint of insulation and isolation is to be underground on an island? Brahe eschewed the social character of places of knowledge production. Solitude was regarded as productive, and it protected the work from contamination—but such levels of isolation are often the subject of suspicion. As Steven Shapin writes, "Solitude is a state that symbolically expresses direct engagement with the sources of knowledge—divine and transcendent or natural and empirical. At the same time, solitude publicly expresses disengagement from society, identified as a set of conventions and concerns which act to corrupt knowledge."[32] Brahe, with his prosthetic metal nose (he'd lost part of his

Figure 3.2
Tycho Brahe's Observatory of Uraniborg (1576–1580). This drawing is from the 1603 edition of Tycho's *De Mundi Aetherei Recentioribus Phaenomenis Liber Secundus* [The Second Book Concerning More Recent Phenomena of the Ethereal World], originally published in 1588. Courtesy Library of Congress/Science Photo Library.

nose in a duel over a disagreement about a mathematical formula), rumored liaison with the queen, and pet elk, was easily as suspicious a character as Nemo.

Atmospheric Islands: Biosphere 2

What Sloterdijk refers to as "atmospheric islands" are the environments of the terrarium, hothouse, greenhouse, and glasshouse. These places recreate the practical conditions that allow the propagation of species and varieties in places whose climates would otherwise never allow them to survive. Noting that the construction of atmospheric islands has come to relate less and less to survival and more to practices of commodification and capitalism, Sloterdijk argues that such human-generated spheres come to constitute "an autonomous, absolute, context-free house, the building with no neighborhood; it embodies the negation of the environment by the artificial construct in exemplary fashion."[33] What Darwin had referred to as the "conditions of life" would become a mere triviality to the contemporary human in Sloterdijk's account.

Sloterdijk's examples of atmospheric islands roam from Joseph Paxton's magnificent "postmodern" Crystal Palace designed for the 1851 Great Exhibition in Hyde Park, London, to the Biosphere 2 project of the 1990s in the Arizona desert.[34] Biosphere 2 is an isolated ecological laboratory, one in which the human subject is part of the experiment.[35] Sloterdijk notes atmospheric islands aren't just "made," they also "make," and the scientist is constructed in Biosphere 2 as a "Biospherian"—an other-world explorer and a television personality, with all the necessary marketing, uniforms, and photo opportunities (figure 3.3).

Construction began on Biosphere 2 in 1987, the same year *Star Trek: The Next Generation* hit our television screens, and was completed in 1991, as Season Four drew to its anti-climactic conclusion. Space, we were told is "the final frontier," and the Biosphere 2 project was to prepare us for life beyond the Earth. The ambitious Biosphere 2 project was financed by Space Biosphere Ventures, headed by the anthropologist, ecologist and engineer John Polk Allen (also known as Johnny Dolphin). Allen's MBA from Harvard Business School helped him propel the Biosphere 2 project into a venture capital media event. Most of the funds came from Edward Perry "Ed" Bass, the petroleum billionaire, who ironically harbored conservationist tendencies. It was Bass and Allen who promoted Biosphere 2 as a "closed-system experiment" in autonomous existence, which

Figure 3.3
Biospherians before the experiment, 1990. Courtesy Library of Congress/Science Photo Library.

was intended to occur between September 26, 1991 and September 26, 1993. Biosphere 2 would house a small population of eight, including scientists, medical doctors, and researchers, at least one future television host, and a dive instructor. Posed in matching uniforms the Biospherians reminded the viewing public more of Mork from Ork than the dashing cast of *Battlestar Galactica*.

The Biosphere 2 enclosure stretched over three acres and contained five "biome" areas, including a rainforest and an "ocean" with a "coral reef." A glass space-frame system covered most of the facility, and all energy was generated on site. This laboratory aspired to be a space station or *Nautilus*, but on the scale of an island. This logic was rehearsed in many science fiction films that posited closed-system domed greenhouse-like structures or space-stations: *Silent Running* (1972), *Solaris* (1972), *Logan's Run* (1976). The Montreal Biosphere (1967), designed by Buckminster Fuller, was itself the site for a 1979 episode of *Battlestar Galactica*. Much like the fraught conclusions to many of these films, Biosphere 2 was not entirely successful. Most of the animals within Biosphere 2 died though the cockroach population reportedly did exceedingly well). As the emaciated scientists ate up their oxygen supplies, it became necessary to pump in air to keep them alive. They bickered among themselves. The laws of the experiment were broken in every breach of the boundary. A medical issue necessitated one of the Biospherians to visit a local doctor, and they were spotted reentering Biosphere 2 carrying plastic bags, the contents of which kept journalists speculating. Isolation and insulation were at once the laboratory's defining feature and its ethico-aesthetic quandary. Every breach of the space-frame boundary that entombed the Biospherians was the subject of the intense media interest that the project had initially furtively encouraged. The death knell of Biosphere 2 may have sounded in 1993 when alt-right activist Steve Bannon was brought in to manage its finances.

Anthropogenic Islands: Janelia Farm

Sloterdijk's third type of island—the anthropogenic island—generates human life. This island is involved in a definitive process of social engineering and production, which occurs in particularly close-formed ways. It is here that we can, after a long journey, return to Janelia Farm and its scientists. Sloterdijk would note that "once a great exaggeration becomes obsolete, swarms of more discreet upsurges arise."[36] In several ways, Janelia Farm is a less-sensationalized—albeit enlarged—version of Biosphere 2. As we already noted, its founders conceived it as an experiment in bringing together scientists from many disciplines into a space without intrusion (insulation) or distraction (isolation). It is without the institutional neighbors and hosts upon which other research centers in the biosciences depend—the universities and hospitals, as well as the commercial and residential precincts that feed, entertain, traditionally house, and socialize scientists. Viñoly's architecture put most of what is deemed necessary for the life of a scientist into the program of a single site. The main building is 1,000 feet (305 meters) long, sinuous in plan, and comprised of a series of low-rise terraces that followed the topography of the site. If stood on its end the building would be the height of an eighty-five-story sky-scraper. The Landscape Building, as it is now called, encloses an area four times larger than Biosphere 2 (581,000 square feet, or 13 acres/5.4 hectares). It shares with Biosphere 2 a great deal of glass and the creation of a new nature. At Janelia Farm the landscape is attached to the outside of the building as a series of planted terrace roofs. The HHMI describe it as both *recreating* the landscape and as a "landscape" laboratory building in itself. Optimistically, it is also described as "an indistinguishable part of the sloping meadow below."[37] But, given the size of the building, we cannot imagine anyone mistaking it for a meadow. When photographed, the new artificial lake created to the north of the laboratory suggests its location as being the foreshore of an island (figure 3.4).

The scale of investment at Janelia Farm is underwritten by a mission not related to solving a big medical issue or developing a pharmaceutical product, but rather given to nurturing the culture of scientific research. It is a focus on "people, not projects."[38] Such a tactic is consistent with the current naming of the fifty-three laboratories of Janelia Farm. Projects here tend to be in neuroscience and range from research into spatial learning in rats and mice to the neurobiology of survival needs such as hunger and thirst. Each laboratory is known by the name of its chief scientist rather than by the project being undertaken or problem being engaged. The eight members of the "Card Lab," named after Gwyneth Card, for example "have identified a set of escape maneuvers performed by a fly when confronted by a looming object."[39] A laboratory that tethers flies and scans their brain activity could have been called "Swat Team,"

Figure 3.4
The laboratory buildings and lake at Janelia Research Campus (2006). Architect: Rafael Viñoly. Photograph by Tuo Peter Li.

but instead the lab remains tethered to its chief scientist. It is this focus on the cultivation of researchers that makes Janelia Farm an atmospheric island and also explicitly anthropogenic. This concentration on people and culture is constituted in particularly troublesome rhetoric, most clearly articulated in *Janelia Farm Research Campus: Report on Program Development* (2003):

> Although there are numerous organizational "cultures" in which scientific research is conducted, from our perspective, no single culture has emerged as "the best." Despite their variety, two factors have had the largest influence in shaping the organizational cultures of research laboratories—the conditions attached to the research funding and the career structures available to the participants.[40]

One doesn't need to be a scientist to recognize the problems with the logic of such a diagnosis. Janelia Farm mobilizes the massive financial resources of the HHMI to provide "a select group of scientists with the facilities, finances, and freedom they need to pursue original, long-term research with minimal distractions."[41] It does this to cure the culture of scientific research, or at least to reimagine a culture that may be "the best."

Janelia Farm does offer minimal distraction. It is a veritable *Nautilus* permanently trapped on the banks of a constructed lake. Once you have arrived at Janelia Farm on the WiFi-enabled shuttle bus service from Arlington, you are well insulated and have no reason to ever be distracted from your scientific pursuits. In addition to the laboratories, Janelia Farm offers: a cafeteria with free coffee twenty-four hours a day; a pub (where alcohol is not free); a library and art gallery; a one-hundred-room hotel for conference visitors, as well as what they call "transient" and long-term housing in a residential village. That sounds lively, but Mitchell Waldrop wrote in *Nature*, it is "not exactly a warm and cosy place."[42] Waldrop invokes descriptors that resonate with Marc Augé's accounts of "non-places." The hallways alongside its laboratories "feel a bit like airport concourses"[43] (figure 3.5). Waldrop finds the curvature of the building's walls "seem to be constantly vanishing around the next bend, which can produce the disconcerting sense that one is stepping off into infinity."[44] And he does not mean this in the positive sense of Buzz Lightyear's catchcry. Another critic coins the term "Jetsonesque" for its architecture of curved glass, white panels, and steel frames.[45] The architecture of Janelia Farm works hard to express a vision of the future, but it is a vision that repeats the problematic abstractions of modernist planning for "new towns" and atmospheric redactions of large-scale commercial architecture.

For Latour laboratories are invariably engaged in practices of insulation that occur as the inverse of physical isolation. Latour's paper "Give Me a Laboratory and I Will Raise the World" traces the path taken by Louis Pasteur, who "situated his laboratory on the farm."[46] This laboratory was at once "in the midst of the world" and still defined

Figure 3.5
The glazed corridor of Janelia Research Campus (2006). Architect: Rafael Viñoly. Photograph by Paul Fetters, courtesy of Janelia Research Campus.

by a system of internal logic that dictated procedures and operations. For Latour, "The macrocosmos is linked to the microcosmos of the laboratory,"[47] and "laboratories are the places where the inside/outside relations are reversed."[48] There is in a laboratory a necessary insulation that comes to make the lab an island of order in a sea of chaotic multiplicity—and yet this same insulation makes the laboratory a law unto itself, an intense disconnect whose walls at once frame and contain (figure 3.6).

There is a close relation between insulation and isolation at Janelia Farm. There is also a close relation between geographic isolation and social alienation here. Janelia Farm frees individuals from the rigors of teaching, grant applications, and performance reviews based on short-term output, which remain characteristic of academia. The world of Janelia Farm is, however, controlled in other manners, and every breach of the boundary that configures the research institution is tightly monitored. For Gerald Ruben, vice president of the HHMI and executive director of Janelia Farm, "Two primary factors operate to shape the 'cultures' in which scientific research is conducted." Both are related to the internal mechanisms by which science has traditionally been constituted and its success measured and rewarded: "The conditions attached to research funding and the career and reward structures available to participants."[49]

Figure 3.6
The laboratory façade of Janelia Research Campus (2006) isolates and insulates, reflects and transmits. Architect: Rafael Viñoly. Photograph by Alyosha Efros.

It is these two factors that Ruben seeks to disrupt. The scientist is the key product of Janelia Farm, and those who are offered positions here are as carefully screened as any biomaterial that enters or exits a laboratory. Rubin states, "The most critical factor for the success of Janelia Farm will be our ability to recruit and nurture scientists who possess not only the scientific talent but also the personality traits and intellectual courage required to engage fully in collaborative and interdisciplinary research that tackles difficult problems."[50] References to "personality traits" and "intellectual courage" remind us of Captain Nemo's own recruitment strategy. Janelia Farm's commitment to letting the scientists get on with the job means the boundary condition that admits scientists is rigorously policed. In the *Harvard Magazine*, Ruben speaks of his "hopes to attract people who crave the backing and the faith in them that you express by giving them a million dollars a year in research funding, ... rather than giving them a salary for life."[51]

As Sloterdijk notes, islands are constructed—"made and not found." However, islands also construct. The anthropogenic island is a removal of parenthesis ... those of god and religious structuring, as well as disciplinarity, coastlines, walls, and anything like hard boundaries. The boundaries of the anthropogenic island are far more discrete. They are observed in policy documents, position descriptions, long driveways, and the configurations of vehicular "drop- off" points. Every protocol and procedure insulates.

And every white coat, every latex glove, every free coffee, and every WiFi-enabled shuttle bus that compels the scientist to keep working, enacts isolation. Together these boundary-defining conditions construct laboratories and scientists. Everyone becomes an island of a kind. Or more than this, everyone becomes multiple islands, sliding almost seamlessly from one to the other, as one moves between the laboratory, the childcare center, and on-site residence. Sloterdijk uses the image of foam to describe the liquid complexity by which cultures are constituted. Rubin would suggest the "cultural objectives of Janelia Farm dictated an unusual design,"[52] but such cultural objectives are already architectural. The insulation and isolation of Janelia Farm is not just *expressed* in the architecture but *constructed* by it: the choice of site; the forceful manipulation of the landscape; the imposition of an architecture at the scale of a Hollywood space station or a Las Vegas casino; and the construction of every village facility on-site. The housing, transient and long-term, the hotel that sits along the artificial lake, the family-friendly pub and the childcare facility anchor the scientists and their families to this island. The loosening of the boundaries to encompass "family life" mean even "home" is no respite from the walls that constitute science and the scientist.

4 Transparency

This chapter is concerned with the transparent walls of laboratory buildings and the ideologies that see architects driven to push "the limits of standard production capabilities" in glazing.[1] The Janelia Research Campus of chapter 3 has no passing traffic, no lay public who might wonder about what takes place within. Nevertheless, the architectural envelope of its Landscape Building is intensely devoted to transparency. Its corridors are lined with unusally clear, custom-made structural glass manufactured in a single factory run in Belgium by the 350-year-old company of Saint-Gobain Glass. The 1,700 wall panels were then laminated, solar-coated, assembled at multiple sites in Europe, Canada, and the United States, performance-tested in Pennsylvania, and, finally, lowered by cable into place. The effort was worth it, for the glass, we are told "serves an even higher function" than providing natural light and visibility: "The glass embodies Janelia's philosophy of transparent, collaborative science."[2] The analogy is a common one and leads, by logical conclusion, to the idea that the greater optical transparency of the glass, then the greater the organizational openness. Thus, at the proposed biomedical laboratory at the University of Basel (2020) by Caruso St. John, the architects have specified for the façade thick clear cast glass to eliminate reflections. So convincing are the many attempts at visibly dissolving the walls of science that some laboratories have become a danger to wildlife. The largest application of Ornilux glass in the United States—a glass developed to deter bird collisions—is for Ennead's Bridge for Laboratory Sciences (2016) for Vassar College, Poughkeepsie, New York.

The South Australian Health and Medical Research Institute

The qualities of the transparent envelopes and membranes—both phenomenal and rhetorical—are explored in this chapter through the South Australian Health and Medical Research Institute (SAHMRI, 2014), and its future neighbor, SAHMRI 2, which is expected to be completed in 2021. The SAHMRI, fittingly, is in the "City of Churches," the moniker for Adelaide popularized by English novelist Anthony Trollope in his 1872 travelogue *Australia and New Zealand*.[3] Both the finished and the proposed buildings are by Woods Bagot, a practice whose first commission, in the nineteenth century, was to extend a cathedral. On a site straddling the railway lines that divide the city's center from its parklands and river, SAHMRI occupies a prominent position. Detached from the urban fabric like a cathedral, it can be seen from all sides. It has responded to this exposed condition with a continuous, curvaceous, and faceted glass façade that wraps even its raised underbelly (figure 4.1). Its internal spaces, too, are defined largely by floor-to-ceiling glazed walls.

The SAHMRI is the city's most architecturally awarded building.[4] It is also one of countless recent buildings for research in the biosciences that boast expansive areas of glazing, claimed by their architects and clients as bringing about social transformation and institutional openness. The distinction of the SAHMRI, however, is that it appears to be a continuously glazed three-dimensional object. If the innovation of the modernist curtain wall was the substitution of the glass eye of the window with a glass face, the SAHMRI replaces the face with a glass body. It is more vitrine, or bubble-like,

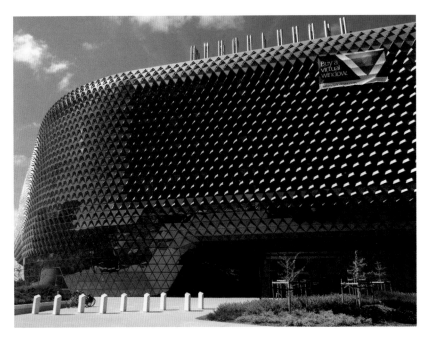

Figure 4.1
Exterior view of the South Australian Health and Medical Research Institute (2014), Adelaide, Australia. Architect: Woods Bagot. Photograph by authors.

than it is like a screen or lens. Glazed building envelopes are a specific materialization of the boundary condition, optically denying the containment they effect. We do not merely observe the action they enclose, we *see through*, and it is this frisson that we find animates both architecture and the scientific pursuit of transparent organisms. In this regard, the SAHMRI shares qualities with two of biology's transparent model organisms, the zebrafish and the nematode. This chapter will explore these similarities in seeking to understand the rhetorical effects of the glass skins of laboratories, not just those of the SAHMRI. We argue transparency gains significance in its engagement with a cultural preoccupation with visibility and its converse, invisibility. This is as true for organisms as it is for buildings.

The 675 researchers at the SAHMRI conduct research in the areas of Aboriginal health, cancer, healthy mothers, babies and children, heart health, infection and immunity, mind and brain, nutrition and metabolism. The institute's 270,000-square-foot (25,000-square-meter) building was funded by the South Australian government and designed by Woods Bagot. In 2017, Woods Bagot were ranked the world's seventh largest global architectural practice with seventeen studios worldwide.[5] The scale and complexity of the SAHMRI are typical of the firm's current portfolio, but the project has special importance as the practice began in Adelaide, making this commission a kind of homecoming.[6]

It was officially opened in November 2014 by the then prime minister of Australia, Tony Abbott, several months after the scientists had occupied it, and cost USD 151 million (AUD 200 million). The following year, it won the forty-ninth annual *R&D Magazine* Laboratory of the Year Award.[7] The award is not insignificant. Eero Saarinen, Mies van der Rohe, Mitchell/Giurgola, and Foster & Partners are past winners. The jury commended the architects for the flexibility of the SAHMRI's eight floors of wet and dry laboratories, its "adequate separation of lab hazards and people movements," and the sensible planning of its support facilities that allow for "service maintenance and changes to occur with limited disruptions to the internal labs."[8] When measured against the pragmatic demands of a highly serviced laboratory, the SAHMRI is very accomplished and benefited from a large team of consultants and the breadth of skills in Woods Bagot's local and international offices.

The response to the building from the architectural and popular press stands in stark contrast to the pragmatic assessment of the *R&D Magazine* jury. That jury was less than interested in the seductive appeal of its curvaceous form and crystalline skin of 6,290 triangular glass panels that transfixed design-savvy commentators (figure 4.2). "Who knew that disco bling could be an ideal motif for a twenty-first century research facility?" writes one admirer.[9] It has been described as a "cut-glass brooch,"[10] a cheese-grater, and an alien spaceship. More frequently, the analogies are drawn from nature. In Designboom, a popular website with content generated by media releases from

Figure 4.2
The articulated glass skin of the SAHMRI (2014). Architect: Woods Bagot. Photograph by Jason Millward.

contributing architects, the SAHMRI is reported to have been "conceived as a living organism."[11] The architects claim that "the form and its articulated skin adapt and respond to its environment."[12] Commentators liken it to various animals and organic structures: a giant shingleback lizard;[13] a two-winged butterfly inside a cocoon;[14] a hive;[15] a prickly pear;[16] and a creature with an already built in armor of self-defense.[17] The metal hoods over the glazed panels are said to be "eyelid-like."[18] The experience of entering the building, according to one journalist, is like being ingested as the "striking exterior becomes a see-through skin, and inside, you're enveloped by the framework of a giant body."[19] Woods Bagot's Enzo Caroscio describes how they designed the columns of the building: "Like a person, they are sculpted to look more skeletal and bone like."[20]

While the architects cite the pinecone as their inspiration, they are not working directly and innocently from "nature"—a construct we interrogate in the following three chapters. The SAHMRI is borne out of a knowing engagement with architectural history and contemporary discourse. Its shape and faceted skin closely resembles that of the Esplanade Theatre (2002) in Singapore—colloquially known as "the Pineapples"and "the Big Durians." The likeness is more than accidental. Atelier 10, the architects of the Esplanade Theatre, were the façade and environmental consultants for the SAHMRI. Working with Atelier 10 and engineers at Aurecon, Woods Bagot engaged current technologies in the design of the building's envelope using parametric modeling tools and environmental performance analysis software. The tools have been honed across a series of buildings with curved diagrid facades, the most well-known of which is Norman Foster's 30 St Mary Axe, London (known as "The Gherkin"). The façade of the SAHMRI is composed of high-performance low E glass, and metallic sunshades are dimensioned according to orientation and the desired levels of light penetration.

Several well-worn themes converge in its design. Zoomorphism and the idea of a "living building" is one. Iconicity is another. It pursues a bold silhouette that stands in detached contrast to the surrounding environment. The theme that interests us most here is transparency. Extensively glazed laboratory buildings like the SAHMRI have obvious precedents in what Peter Sloterdijk called "atmospheric islands," including the glass buildings associated with science, such as the greenhouse and hothouse. Specific examples include the Crystal Palace (1851), Bruno Taut's Glass Pavilion (1914) (figure 4.3) for

Figure 4.3
Glasshouse Pavilion, Cologne Werkbund 1914 Exhibition. Architect: Bruno Taut.

the Cologne Deutscher Werkbund Exhibition, and Biosphere 2 (1991). The physical qualities that make glass attractive for laboratory buildings (and for laboratory equipment) are multiple: it is sterile, impermeable, hygienic, and transparent. Glass can be manufactured to different hues, thicknesses, sheet sizes, and shapes and imbued with varying degrees of transparency.

Transparency is a measure of the physical property of light passing through a material without being scattered, so objects or images can be seen as if there were no intervening material. It is an empirical quality that one would imagine to be unproblematic. But in practice, glass in its architectural application never entirely disappears. This is despite such technical innovations as glass fins and beams to eliminate opaque structural members. Even when it is not tinted, fritted, etched, or reflective, the material presence of glass is given away by the tell-tale signs of joints, cracks, dust, condensation, rivulets of rain, the residue of window cleaning fluids, fingerprints, or guano. Worse, a pane of glass is revealed by a smear marking the spot where, for a bird, the sky unexpectedly solidified.

Glass and Its Associations

The meanings ascribed to glass are unstable and contradictory. Its optical transparency is often conflated with informational accountability, although one does not necessarily follow the other. Since the 1950s many jurisdictions in democratic societies have established open information acts that compel public institutions to "transparency." In business, government and law, "transparency" refers to the full disclosure by organizations and individuals of their fiscal operations, decision-making, policies, and records. It signals free access to any information of interest to employees and the public. Scientific research organizations have been under pressure since the 1970s to be more transparent in this sense of the word. Diminished confidence in scientists and science led to changes in the behavior of scientific institutions and how they are represented rhetorically and architecturally. Changes in funding, too, have meant that even scientific organizations in universities have come to depend on public support and private philanthropy, for which secrecy, invisibility, and remoteness may be an obstacle. The need to convey a commitment to informational transparency has, ostensibly, become a strong architectural driver, with extensive glazing the go-to technique used by scientific organizations to manage the contradictory needs of security and apparent openness.

The glazed laboratory building has been acclaimed by architects and scientists because of its alleged capacity to signify institutional openness. It has also been celebrated on the basis that internal transparency will bring about social transformation. At the SAHMRI, transparency allows "outside views to the internal workings of the building promoting the importance of the activities within"[21] (figure 4.4). Such statements are commonly made by those who commission, design, critique, and come to occupy laboratories. The architects of the University of Oxford's New Biochemistry building (2008), Hawkins Brown, similarly state, "All of the elevations are transparent ... mak[ing] a statement about the value and integrity of the biomedical research inside."[22] Sometimes, however, the logic of transparency and visibility becomes confused. Mikkelsen Architects write of the Laboratory and Logistics Building (2018), Copenhagen, Denmark, "It is our intention to make the building's various activities transparent, as well as being visible from outside."[23] At the Sainsbury Laboratory, University of Cambridge (2011), there are "glazed walls which let other people see what's going on."[24] Of the Frick Chemistry Laboratory, Princeton University (2010), it was said, "The scientists are now going to be working in this remarkable glass loft, and they're going to have views through the building, but always with a sense of openness."[25] More vivid still, is the apiarian metaphor used by one critic to describe the Allen Institute (2015) for brain and cell science in Seattle, Washington, designed by Perkins and Will: "The design orients labs like flower petals around a large light-filled central atrium; the effect is like the inside of a bee hive where researchers can see each other and what they are doing, making the space more collaborative, flexible, and transparent."[26] The metaphor recalls Jeremy Bentham's description of the Panopticon: "It will be a lantern; it will be a beehive; it will be a glass bee-hive, and a bee-hive without a drone."[27]

Glazing, as an instrument by which to influence the behavior of scientists and convey ideological aspirations, is a construct rarely challenged by contemporary architects or their clients. Glass has come to *equal* visibility, which *equals* sociality, community, trust, knowledge-sharing, openness, and so on. It's a metaphor that goes back to Jean-Jacques Rousseau, who imagined French post-revolutionary society as one without misunderstandings or injustice. Rousseau uses the image of the glass heart, "transparent as a crystal," to conjure a vision of a society where there are no secrets between citizens, or citizens and the state.[28] Engraved on Taut's Glass Pavilion were maxims from Taut's friend, the poet Paul Scheerbart, that echo Rousseau: "Without a glass palace life is a burden," and "Colored glass destroys hatred."[29] The Constructivists thought a transparent building would destroy the distinction between the private and the public and that the application of glass in modern architecture would herald a new culture in which the shadows of the past would vanish and secrecy in the present would be impossible.

The glass building as a vehicle of redemption and social transformation gained momentum across the twentieth century. Walter Benjamin, in his 1929 essay on Surrealism extolled the revolutionary attributes of glass architecture: "To live in a glass house

Figure 4.4
The foyer of the SAHMRI (2014) reveals nothing of the science within. Architect: Woods Bagot.
Photograph by authors.

is a revolutionary virtue par excellence. It is also an intoxication, a moral exhibitionism that we badly need. Discretion concerning one's own existence, once an aristocratic virtue, has become more and more an affair of petit bourgeois parvenus."[30] Benjamin suggests that glass is "the enemy of secrets. It is also the enemy of possession."[31] In "Experience and Poverty" (1929), Benjamin commented that through glass architecture, "Everything comes to stand under the banner of transparency."[32] While director of the Bauhaus during 1928–1930, Hannes Meyer concurred. He believed glass would destroy corruption through "open glazed rooms for the public negotiations of honest men."[33] Glass was, thus, vested with political connotations as an agent for honesty and representative democracy. What the media, the "fourth estate," were to accomplish through disclosure, glass was to effect in the spatial realm of the city.

The rhetoric around glazing at the SAHMRI and other glazed laboratory buildings draws from this rich history, but there is an alternative lineage linking architectural transparency with unwanted observation. The same techniques, practices, and policies intended to ensure transparency in business and politics are equally put to work to enable intrusive surveillance and control. Jacques Derrida linked totalitarian societies with the mantra of transparency in political institutions, declaring that "if a right to the secret is not maintained, we are in a totalitarian space."[34] And Nigel Whiteley noted something of a revival of glass architecture due to public and social redefinitions of "transparency." He cautions that what is taken for transparent processes in society are often, in fact processes of surveillance and scrutiny, marketing and spectacle.[35] This link between glass architecture, surveillance, and control has a longer history.

The Crystal Palace

In 1863, Charles Darwin, who had several unheated glasshouses in his gardens at Down House, decided he needed a hothouse. He had just published *The Various Contrivances by which British and Foreign Orchids Are Fertilized by Insects* (1862), his first investigations into plant fertility. In order to expand and clarify his theory of evolution, to refute his detractors, and to better understand the mechanisms of species survival and reproduction, Darwin needed glasshouses in which he could create artificial islands of varied heat and humidity. In them he would grow orchids that had evolved to attract just one species of insect and carnivorous pitcher plants that sensed and responded to the presence of their prey. The plants, with their animal-like traits and interactions, suggested links between life's two kingdoms and were essential to the hypothesis of a common evolutionary origin to all organisms. Also, there may have been more frivolous reasons for the glasshouses. Darwin wrote to his friend Joseph Hooker, then deputy director of

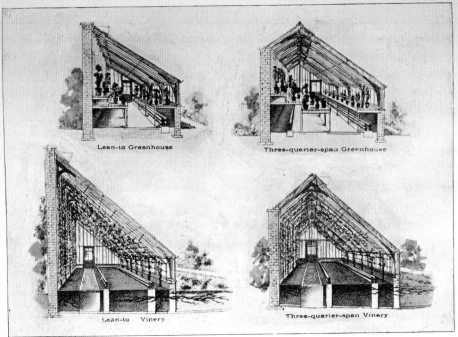

Figures 4.5a and 4.5b
Sections of glasshouses available from hothouse builders Mackenzie and Moncur (1892 catalogue; repr. Edinburgh, 1901).

the Royal Botanic Gardens at Kew and a regular supplier of his specimens, saying the stove-plants "do so amuse me."[36]

Darwin's glasshouses and the Royal Botanic Gardens signaled an epistemological shift, but they also relied heavily on new technologies, and glass technologies, in particular. First came the invention of wrought-iron glazing bars in 1816, then Nathaniel Ward's 1829 design of a sealed glass case that allowed the transportation of live plants on sailing ships. Next was the abolition in England of the Glass Excise Tax in 1845, which was followed two years later with James Hartley's patenting of industrially produced rolled plate glass. The Wardian case, which also played a part in Darwin's story, was a direct forerunner of the terrarium, vivarium, and aquarium, glazed containers each playing a significant part in laboratory science. It had become impossible to grow plants in the gloom of industrialized cities, and this intensified the attractiveness of glasshouses. The network of railways in the UK, which doubled between 1850 and 1868, allowed garden and building materials, as well new-fangled gadgets such as lawn-mowers and gas lights, to be transported quickly around the nation. The institutional support and enthusiasm for botany saw the Royal Botanic Gardens richly resourced and backed up by an integrated system of colonial gardens and overseas university departments. The same enthusiasm launched a publishing industry centered on botany and gardening. These factors seeded and nourished a craze for gardening and exotic plants, which saw the glasshouse emerge as a quintessential architectural innovation of the Victorian era (figure 4.5). An archipelago of florid "atmospheric islands" shrouded in the smog of industrial Britain had bloomed. The horticultural glasshouse, Dustin Valen argues, were not only laboratories for studying organic life; they "provided an important setting where the triumph of science over nature was put on full display and where new environmental technologies could be developed and tested."[37] In the long term, the success scientists and horticulturalists had with glasshouses shifted architectural attention toward questions of environmental control, indoor climate, and hygiene, but architects were at first wary.

The largest and most influential of these island-cum-cathedrals was the Crystal Palace, constructed in 1851 for the Great Exhibition in Hyde Park to the designs of horticulturalist Joseph Paxton. Its lofty arcade, under which whole trees thrived, inspired many amateur gardeners to the acquisition of their own greenhouses. Paxton capitalized on his new fame to launch a range of affordable, modular greenhouse kits. Architects and aesthetes were less than impressed, not the least because Paxton's design had been selected over 245 other schemes, many of which had been produced by eminent architects. The Crystal Palace sat outside the classical canon. Detractors thought its pedigree, as well as that of its designer, were dubious. The art critic John Ruskin damned it as a "great cucumber frame between two chimneys."[38] Augustus Pugin decried it as "a glass monster."[39] Architect George Gilbert Scott called it a "miserable travesty" and sarcastically asked whether it is more like a Gothic cathedral.[40] The *Times* referred to the Crystal Palace as a "monstrous greenhouse" and campaigned against its construction. Feodor Dostoevsky in *Notes from the Underground* (1864) declared it a "glass chicken coop." Whereas Ruskin's objections were snobbish, Dostoevsky was deeply affected by what he saw in the Crystal Palace as the harbinger to a rational, materialist, and mathematically precise utopia.[41]

Through a Glass, Darkly

Echoes of Dostoevsky are heard over half a century later.[42] Yevyegny Zamyatin's novel *We*, an indictment of the police state that the Soviet Union had become, was smuggled out of the country to be published in English translation in 1924.[43] *We* opens with the imminent completion of "the glass, electric, fire-breathing Integral." The Integral is a space ship of durable and flexible glass that will bring a "mathematically faultless happiness" to beings on other planets still burdened with the primitive state of freedom.[44] To be launched from the United States, the Integral houses a city composed of glass and bounded by a Green Wall. While beautiful, with its "impeccably straight streets," "glistening glass pavements," and "round bubbles of cupolas," such a city would also be a place of complete surveillance, Zamyatin predicts. Its inhabitants would be devoid of individuality, and their habits, movements, and reproductive relations strictly regulated. When the protagonist D-503 awakens to the morning bell, he sees "to the right and to the left through the glass walls ... others like myself, other rooms like my own, other clothes like my own, movements like mine, duplicated thousands of times."[45] He is awed by the elegance of this machine and the architecture of the city, where

> on days like these, the whole world is blown from the same shatterproof, everlasting glass as the glass of the Green Wall and of all our structures. On days like these, you can see to the very blue depths of things, to their unknown surfaces, those marvelous expressions of mathematical equality—which exist in even the most usual and everyday objects.[46]

While admiring the precise beauty of this united body, D-503's new and unexpected lust for citizen I-330 unleashes disturbing dreams. He discovers he has a soul, a feeling he likens to "some foreign body like an eyelash in the eye."[47] As he struggles with new insights and desires that are more personal than collective, the consequences and contradictions of this utopia are revealed to the reader. Although transparent walls ensure the city's citizens "have nothing to conceal or to be ashamed of," and elections are held openly, D-503 recognizes the situation makes the task of governing "much more expedient" for the unseen Guardians.[48]

The denizens of this city are also metaphorically transparent and, at times, literally so. When D-503 meets his state-approved consort, O-90, at the arranged hour, he marvels at how "her round blue eyes opened toward me widely, blue windows leading inside; I penetrate there unhindered; there is nothing in there, I mean nothing foreign, nothing superfluous."[49] A traitor, awaiting public execution, is observed to be "already colorless, glass face, lips of glass."[50] After the execution, D-503 feels "purified and distilled, transparent."[51] His tranquility is destroyed later the same day by the poison of illicit liqueur (or the lust inflamed by a kiss from which he takes the poison). He says, "I became glass-like and saw within myself. There were two selves in me."[52] Self-knowledge comes with the recognition that under the glass city "something wild, something red and hairy, was silently seething."[53] What, we might ask, is silently seething behind today's glass cities and buildings?

Glasshouses, Glass Cities

Several decades after the publication of *We*, glass cities are everywhere, from Houston to London to Hong Kong, but as Dostoevsky had warned in *Notes from Underground*, they are "terribly boring, for everything [is] rational and calculated."[54] The uptake of transparent curtain walls in office blocks of the post–World War II period had led, as Charles Jencks remarks, to the "contradiction between the technical and visual excellence" of the transparent curtain wall on the one hand and "the undeniable banality of the building task on the other."[55] The problem was not with glass per se, but with a regulated life of conformism and tedium that played out in the standardized spaces within. The glazed tower or corporate office, as much for what it houses as its efficient architecture, is a capitalist success story—of anonymous, serial, modular, abstract, and technical systems. Reinholdt Martin argues that the reproducibility, modularity, and flexibility implied by the glazed curtain wall "became the image—and the instrument—of the organizational complex," of the military and industrial sectors.[56] Martin describes the commercial buildings dominating the landscape between 1950 and 1980 as "empty skins full of individualized consumer-subjects that, like giant television sets, organize through the agency of an auratic delirium."[57]

The relationship between control and consumption, material transparency and political opacity continues. It is perfectly obvious, even to those without architectural training, that laboratory buildings are heavily securitized, and public access tightly controlled. It is obvious, too, that what can be glimpsed of laboratory work from foyers or from outside is only part of the story, and in any case, performed in a language of gestures and actions few understand. Bruno Latour argues:

> Demanding that scientists tell the truth directly, with no laboratory, no instruments, no equipment, no processing of data, no writing of articles, no conferences or debates, at once, extemporaneously, naked, for all to see, without stammering nor babbling, would be senseless.... The direct, the transparent and the immediate suit neither complex scientific assemblages nor tricky constructions of political talk.... If we start making direct and transparent processes the supreme law of any progress, then all scientists are liars and manipulators, and all politicians corrupt bastards.[58]

Latour's point is not so much about glazed buildings, but about the complex ecology in which scientific knowledge is produced and how that necessitates texts and infrastructure inaccessible to the populace. He adds that science gains its power by collecting inscriptions, and other things, and moving them to a distant center where they are assembled, compared, or contrasted to produce knowledge. Paradoxically, scientific knowledge depends on the "referential chains" between scientific texts, as well as their separation from public discourse and oversight.[59] The laboratory is a key link in these chains. The separation of the laboratory from public view and comprehension is not just about the communication of facts; it also entails one's moral responsibility to those facts. The division between laboratory and world persists after knowledge is produced. While "deep in your laboratory you can revolutionize the world," writes Latour, on the "other side, others will suddenly have to take care of the consequences—ethical, political, and economic."[60] This asymmetry of action, information, and responsibility imbues science with authority. The glazed building maintains this asymmetry. We see, but we do not comprehend, nor can we intervene. We are reassured by a gesture of openness, despite all parties knowing there have been no secrets divulged. In fact, nothing is revealed. Or more precisely, the transparent building implies *there is nothing to reveal*.

In this regard, while the rhetoric around scientific transparency may be facile, the transparent building is also productive, in that it "makes a spectacle of the scientific work."[61] This phrase comes from a journalist responding to the University of Pennsylvania's Krishna P. Singh Center for Nanotechnology (2013) designed by Weiss Manfredi. He elaborates: "Scientists can be seen from the courtyard through the structure's three layers of glass, where they appear behind a wide amber glass partition ambling about like extras in a science fiction film.... Here, the architects tinted entire glass walls, rendering the lab space shockingly transparent."[62] The scene/seen distracts, entertains, and appears to us in cinematic fashion. In framing the scene like a screen, or a proscenium, the glass wall creates distance. It separates. It offers us an image, such as those Derrida finds in scenes "that violently fill the view or rush the mind's eye," which is something akin to "the blindness necessary for theater."[63] There is a backstage for every laboratory that is not disclosed by the transparency of its envelope. Of course, scientists also appear to each other as Zemyatin's protagonist and his fellow citizens do upon waking each morning. This intra-surveillance does not start with the scientists; rather, it extends from the centrality of observation in the experimental method. Scientific discovery is propelled by technologies of observation, visualization, and image-making, which reveal the hidden, thereby allowing them to be recorded and analyzed. Even with such developments, the transparent organism, the self-revealing subject, remains a recurring fantasy of science and science fiction.

See-Through Organisms

In *A Guinea Pig's History of Biology* (2007), Jim Endersby demonstrates how the emergence of model organisms depends on a whole system of interactions between the animal; the researchers and their institutional affiliations and friendships; their laboratories and instruments; laboratory technicians; funding; and a myriad of other factors that seemingly lie well outside science. Model organisms are also revealing of the values and experimental interests of scientists at different points in history. For Darwin's contemporaries, it was barnacles and pigeons. In contemporary popular culture, experimental science is associated with rodents. Other "model organisms" have emerged over the past century: the fruit fly, the bacteriophage, the nematode, and the zebrafish. What seethes silently behind the glass walls of the laboratory, and in its glass dishes, is not "red and hairy" but transparent.

Transparency is a quality of certain mutations and species of nematode and zebrafish. It is a quality that has significance only in the laboratory context under the gaze of the scientists and is of no use to the creature in the wild. For Sydney Brenner, finding a transparent animal was not his end goal;[64] instead, "the superior optical qualities" of *C. elegans* was an "unforeseen advantage."[65] It made it possible for investigators to "actually watch the process of development unfold in a living animal under a microscope."[66] Andrew Brown, author of *In the Beginning Was the Worm* (2003), describes the nematode as "a transparent lens through which the rest of biology can be studied."[67] Science writer Leslie Roberts enthuses, "You can look at a complete neural circuit for a particular behavior and get a complete and convincing description of the nature of that behavior.... You can look at it and say 'that is all there is.'"[68] In 1998, the nematode became the first multicellular organism for which full genome sequencing was completed. But its elemental form constrains the nematode's utility as a model for understanding and, ultimately, intervening in the human body—the end game of biological research.[69] Hence, the superiority of the zebrafish for medical research.

The introduction of the zebrafish to the laboratory came via the nineteenth century's appetite for exotic species in domestic aquaria. It arrived from India in Europe in the early 1900s and was then introduced to the laboratory in the 1930s by Charles Creaser. Creaser's advocacy of the fish was helped by the serendipitous fact that zebrafish ate *Drosophilia*, the wingless fruit fly already abundant in laboratories. The zebrafish (*Danio rerio*), an invertebrate, became an important model organism for vertebrate development, genetics, and human biology and disease.[70] The transparency of the wild zebrafish, however, is confined to its infancy, or at least it *was*. In 2008, a team of scientists in Boston "created" a transparent adult. The transparent zebrafish, named the *casper* for its ghost-like appearance, was bred using two mutant lines, the *nacre* and the *roy orbison* (a line with unusually large black eyes). The transparent adult zebrafish of the *casper* line offers "a unique combination of high resolution, sensitivity and amenability to deep tissue imaging with commonly available laboratory equipment."[71] Within a year more than a hundred laboratories had adopted the *casper* to study cancer and pathology (figure 4.6).

The first use of time-lapse photography in biology, by Warren Harmon Lewis, had as its subject the transparent eggs and embryos of the zebrafish. The spectacle of its growth was astounding, for Lewis could speed up on the screen "activities far too slow to be comprehended by direct vision."[72] Technologies for "seeing" inside living organisms, from time-lapse photography to x-rays, ultrasound scans, and magnetic resonance imaging, have been central to modern biology and, theoretically at least, should have made physical transparency to the naked eye redundant, but this is not the case. For Richard Doyle, the transparent worm is the sublime object of biology (his book was released before the announcement of the transparent zebrafish), because it is "the continual story that there is nothing more to say, a story of resolution told in higher and higher resolution."[73] A decade later, Melinda Wenner reported in *Nature Medicine*, "Biomedicine would be a breeze if organisms were transparent.... Biologists could study exactly how an animal's organs develop by observing them as they grow. In effect, the secrets of the body would be out there for everyone to see."[74] She reports, researchers "are even making strides toward turning human tissue transparent," with one scientist describing his fantasy of a human like a jellyfish.[75] The desire to see and know all rejects the limits of perception dictated by the human eye. There is pleasure in folding back time, exposing the tiny, bringing the distant into view, and seeing through the skins of bodies. There is a beauty and terror in making the invisible visible, and the visible invisible.

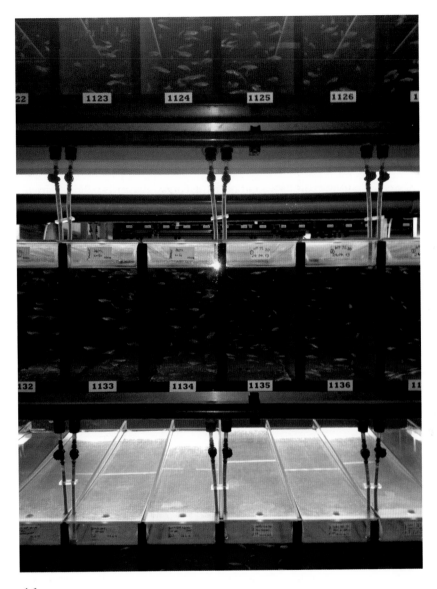

Figure 4.6
Zebrafish facility in a biosciences laboratory. Photograph by authors.

Skins

From the Riviera in Adelaide (an optimistic name for a mediocre hotel in the Comfort chain), it is possible at dusk to spy one of the SAHMRI's scientists (figure 4.7). Our voyeuristic gaze penetrates the room's windows, crosses the street, and continues through the SAHMRI's glazed perimeter. The laboratories, like the office of the scientist, are enclosed floor to ceiling by glass. The laboratories in each wing of the building are connected by a glazed and enclosed bridge that has biological containment certification. These bridges narrow at their center, creating a sense of tension, as if the wings of the SAHMRI were straining apart. Here, curves are used to enhance the idea of an organism whose skin stretches and folds and continues through the body in a series of involutions. From the bridge, the scientist can see down into the foyer, and then beyond to the city. Back in her office, she watches a technician using a confocal microscope to better capture the dynamics of a cellular event inside a live zebrafish. The tiny creature

Figure 4.7
Illuminated exterior of the SAHMRI (2014). Architect: Woods Bagot. Photograph by Michael Coghlan.

is swimming in a minute pond in a glass petri dish. We imagine the zebrafish returning its gaze and the series of glances unfolding back toward us, an idea often pursued in arthouse films.

This *casper* fish is as transparent as Zemyatin's traitor as she awaits her fate. Indeed, the scene recalls a room in *We*: "a glass room, filled with golden fog; shelves of glass, colored bottles, jars, electric wires, bluish sparks in tubes."[76] There is undeniable frisson, for us and for the scientist, in these successive acts of looking not just *at*, but *through*: the building's glass envelope; the glazed office; the perimeter of the laboratory proper; the clear tanks of the zebrafish aquaria; the lens of the microscope; the skin of the mutant zebrafish; and the transparent cells mutating and multiplying within. Each pellucid container lies within another. It is not so much the foam imagined by Sloterdijk, as the infinite regress of a mirrored room, a glass matryoshka doll. A crystalline series of desire, deferral, distancing, and detachment. Each glass membrane simultaneously detaches and attaches the inside from the outside. You can see but you cannot touch. You do not breathe the same air nor occupy the same space. The fact of this separation, coupled with its visual denial, enthralls. For science historian Jim Endersby the experience of looking through a microscope at the tiny embryos of a zebrafish makes him ecstatic. He says, "You can see right through them, watch their tiny hearts beat and follow the blood as it moves through their perfect glassy bodies. Watching the fish, I felt I understood the fascination of biology more clearly than I have ever done before."[77] The transparent façade of the SAHMRI intends a parallel experience: to dazzle with its glassy body. In the SAHMRI there are no historical or cultural references arranged on the surface, no images drawn from science, no registration of authorial intent, for here science needs no mediation. We *look through* as if there were nothing between inside and outside. The building's glassy logic follows a dream of presentation whereby science makes itself known, without architectural expression. Curiously, this dream of self-effacement has recurred in architecture. In the morning, with the sun glinting against its skin, the dream is over. The SAHMRI has become opaque, reflective, jewel-like. Now its scientists look out at the city, each other, and the fish still swimming in its dish.

Architectural renderings of the SAHMRI 2 suggest the desire to overcome the fleetingness of the dream (figures 4.8a and 4.8b). Its sun-shading system is less "eyelid" than plucked eyebrow; its curtain wall smoother, less reflective. Its client, Commercial & General, claims that it will "complement and accentuate the striking geometric façade of the existing SAHMRI," but it is "still under design."[78] That is, the SAMHRI 2 currently exists only as an architectural rendering, and one that allows architects all of

Figures 4.8a and 4.8b
Conceptual representations of the SAHMRI 2, designed by Woods Bagot, expected to be completed in 2021. Courtesy of the clients, Commercial & General.

the possibilities of transparency. Not the realities but the possibilities. Renderings allow glare, reflectivity, and transparency to coexist in ways that defy optics. The rendering is a visual translation of the rhetoric that surrounds transparency: translating that which cannot be seen into the realm of the visible. Like the MRI, this visibility is about operability. Seeing the organs of the fruit fly allows a scientist to advance a model of understanding, just as the architectural rendering might smooth the waters of planning and development consent.

5 Unbounded

Contemporary laboratory buildings, as we saw in the previous chapter, are often clad extensively in glass to spectacular effect as their architects and clients seek to fulfil complex ambitions spanning the technical to the rhetorical. Floor-to-ceiling glass walls are also widely used inside biosciences research buildings, especially to contain the wet laboratory from the spaces in which scientists write, socialize, lunch, and play. Here, too, glass negotiates the need for biological containment and security against the opposing desire for scientists to move easily around the building, making connections with each other. The border of the wet laboratory does more, however, than control who enters the space. It enables precise and distinct atmospheric conditions—temperature, number of air changes per hour, humidity—that protect the experiment. It contains waste and hazards to meet the demands of physical containment certification.

Fabrikstraße 22 (2010), the laboratory building for the Novartis Institutes for Bio-Medical Research (NIBR) at St. Johann in Basel, rejects this compromise. It does away with the conventional glazed wall in a novel negotiation of the competing aspirations for continuity and containment. Also known as WSJ-355 (like a character in Yevgeny Zamyatin's *We* [1924]), this building designed by David Chipperfield has no walls between the laboratory benches and write-up spaces. Biological containment is invisibly managed through the flow of air and the observance of protocols. Novartis took calculated risks in pursuing this unusual mode of managing biological containment. This chapter contextualizes those risks and their effects within the broader shift toward simulated or "numerical experiments" and the advent of experiments undertaken cooperatively between remote laboratories.

Bruno Latour contests the assumption *"that science stops or begins at the laboratory walls."*[1] For him "the laboratory is a much trickier object than that, it is a much more efficient transformer of forces than that."[2] Scientific facts are made within laboratories, but to make them circulate there must be networks and universal standards that allow experiments to be extended to problems in factories and hospitals, or converted into statistics and policies. Society itself must be transformed into a "vast laboratory."[3] In this book we repeatedly confront ideas about the socialization of scientists through their co-location in settings that encourage connectivity. Yet, conversations between scientists are not confined to the buildings in which laboratory teams are co-located. Conferences and symposia, workshops, printed publications and letters have also been significant for the communication of new discoveries and are now augmented by electronic communications. Indeed, new technologies have made it possible for researchers to collaborate with scientists in laboratories across the world. They have also allowed scientists to control some aspects of experimental manipulation remotely, thereby doing away with some of the need to be physically situated in the laboratory. Following the digitalization of communication and experimentation, Fabrikstraße 22 suggests that a metamorphosis toward a new spatial form might be taking place, one that complicates preconceptions about what is interior and exterior to the experiment. If developments in the design and use of the laboratory register changes in the construction and dissemination of scientific knowledge, then this chapter enquires into the future heralded by the elimination of the walls between the laboratory and its outside—that is, between the experiment and the uncontrolled world beyond.

The Prevailing Model: The Defended Laboratory

The conventional division between the laboratory proper and the spaces of discussion and writing is mirrored in the bifurcations of architectural labor between laboratory planners and design architects. Specialized architectural practices are commonly employed to ensure the functionality of the internal wet laboratory proper. High-profile design architects are commissioned to design the surrounding "rhetorical" spaces of research buildings—expressing in images, spaces, and surfaces the conversational and convivial aspirations of the scientific organization. These architects are also responsible for the external expression, the façades and the public faces of the research facilities. For example, Will Alsop teamed up with Amec on the Blizard Building for Queen Mary University in Whitechapel, London. Gruner AG undertook the laboratory and general planning of Actelion Pharmaceutical's laboratories (2010) in Basel, Switzerland, for which Herzog and de Meuron are the named architects. Princeton University's Frick Chemistry Laboratory (2012) was designed by the British firm of Hopkins Architects with architects Payette of Boston and Jacobs Consultancy undertaking the laboratory planning. Cannon Design planned the laboratories, undertook the engineering, and are the architects of record for Novartis's two new laboratories (2017) in Cambridge, Massachusetts, one designed by Maya Lin, its neighbor by Toshiko Mori.

Numerous laboratory planning consultancies were founded in the 1970s and 1980s. Some, such as Health, Education and Research Associates (HERA, Inc.), were borne out of architectural practices or engineering firms; others were set up by individuals with backgrounds in science and laboratory management. Around the same time, several large architectural and engineering practices developed significant in-house departments in laboratory design. For example: KlingStubbins (since acquired by Jacobs Engineering Group); Payette; Perkins and Will; and Bohlin Cywinski Jackson. In 1984, Research Facilities Design (RFD) was established with the singular mission of providing specialized laboratory planning services. This points to the severance of the expressive aspects of the design of contemporary laboratories from the technical problem-solving, servicing, and planning of wet laboratories and associated equipment. RFD has worked with Foster and Partners on several projects including the Sir Alexander Fleming Building in London (1998) and the Center for Clinical Science Research at Stanford University (1995). With Skidmore, Owings and Merrill LLP (SOM), RFD undertook the Genome Sciences Building for the University of North Carolina (2012). RFD also worked with Woods Bagot on SAHMRI and with Rafael Viñoly at the Janelia Research Campus, the subjects of chapters 3 and 4, respectively. At Janelia, additional consultation was given by Robert McGhee, the architect and senior facilities officer of the Howard Hughes Medical Institute (HHMI). The tension between socialization and containment, pragmatism and expression is perfectly captured by the HHMI's description of the working relationship as a kind of battleground. Campus director Gerry Rubin, recalls that "our trustees felt very comfortable hiring a visionary architect because we had Bob McGhee, who is an expert on lab design. He could push back to make sure the building would meet its scientific function."[4] McGhee led HHMI's laboratory construction and renovation projects for twenty-two years and was responsible for the selection of Viñoly as the project architect. The envelope is a détente between two opposing forces; it is where the pragmatic responsibilities of the laboratory planner meet the theatrical ambitions assigned to the design architect.

This professional assignation of design responsibilities reflects a division of scientific labor and architectural investment, as well as a spatial division—although the territorial demarcations are often as troubled and complex as the glazed walls that contain laboratories or wrap buildings. As the meeting place of two different conditions the glazed wall is a paradox, not unlike that observed by Leonardo da Vinci when he puzzled over the surface of water. The philosopher Avrum Stroll quotes da Vinci in his book *Surfaces* (1988):

> The surface of the water does not form part of the water nor consequently does it form part of the atmosphere, nor are any other bodies interposed between them. What is it, therefore, that divides the atmosphere from the water? It is necessary that there should be a common

Figure 5.1
Fiona Hall, *Out of Mind*, artwork between the foyer and laboratories of the Queensland Brain Institute (2007). Architect: John Wardle Architects. Photograph by authors.

boundary which is neither air nor water but is without substance…. A surface is the common boundary of these two bodies.[5]

Stroll concludes that for this reason surfaces are boundaries or limits and "invested with systematic and profound ambiguities."[6] Glazed walls face both the laboratory and its outside, but belong to neither. These envelopes, which secure biological contaminants to an interior, also operate as a site of transference between interiors and exteriors. They express the activities, ideas, and ideologies of interiors to exteriors, and vice-versa. The glazed envelope operates as a frontier, or in Latour's words, as a "transformer of forces."[7]

The self-effacing quality of the envelope's materiality, its transparency and continuity, sets out to visually discount the physical border it enacts between two very different modalities of scientific labor. On the laboratory side there are the washable, chemically resistant hard surfaces in neutral tones, task lighting, and regularly placed laboratory benches, all of which establish an atmosphere of clinical efficiency. On the other side can be seen the architects' efforts to make convivial workspaces with bright colors and soft furniture arranged for conversation, carpeting, unpainted timber, and softer lighting. Far from a *gesumtkunstwerk*, the research building is often deeply split by these internal contradictions played out in material and appearance. This is why we so often see these internal glazed walls embellished with colored glass, decals, and works of art that mediate and soften the contrast. Fiona Hall's four-story high "Out of Mind" photo collage—articulating the boundary between the neuroscience laboratories and public foyer at the Queensland Brain Institute (2007)—is a case in point (figure 5.1). With its merger of imagery from fruit fly and human brain scans, her artwork both expresses content and discretely obscures the public's view of the scientists. Attention is drawn away from the division of space, to the articulation of surface.

Fabrikstraße 22 (2010)

The Fabrikstraße 22 laboratory exemplifies an astute negotiation of this split. The building takes the form of a simple cube (figures 5.2a, 5.2b, and 5.2c). It is approximately 178 feet (54 meters) long, 100 feet (30 meters) wide, and 5 stories high. It has a floor area of 125,000 square feet (11,600 square meters). The façades are gridded by the horizontal lines of its expressed floor slabs and the vertical pillars, which are spaced 9 feet (2.72 meters) apart. Only two construction materials are visible, glass and an almost white reinforced concrete. What Paolo Fumagalli describes as "the almost dogmatic rigor of the design" is relieved by the introduction of a slight modulation, a chamfer on one side of each of the pillars that is applied seemingly without an overriding pattern.[8] It is not an architecture that anticipates daring departure from convention.

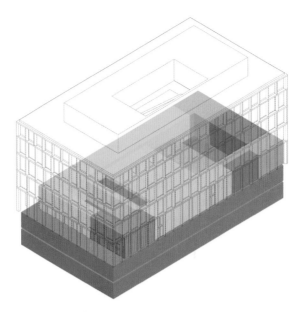

Figure 5.2a

Volumetric Diagrams of Fabrikstraße 22 (2010), Basel. Architect: David Chipperfield. Figure 5.2a shows the ground floor and two basement floors. Support facilities are in brown and social spaces in yellow. Here the main social space is the Cha Cha Thai restaurant.

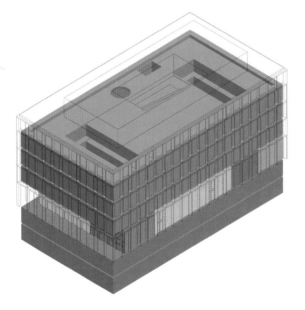

Figure 5.2b

The shared laboratory and office floors in blue. The enclosed stair is spaces are shown in pink and the two lift cores as a void.

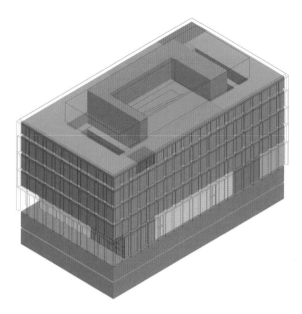

Figure 5.2c
The top floor consisting of offices, garden, and service plant.

Chipperfield dislikes novelty and admits, "I don't think architecture is radical."[9] Despite—or perhaps because of—its unelaborated and subdued exterior, Fabrikstraße 22 presents an exceptional departure in the containment and envelopment of the laboratory function. The elimination of the glass walls around the Fabrikstraße 22 laboratory spaces and its functional replacement by systems of air control is novel. Novartis describes the arrangement in terms of maximizing the interaction between scientists, but we question whether there is more going on. Might Fabrikstraße 22 register the diminishing reliance on the wet laboratory as virtual experimental models of greater power emerge? Could it also suggest that the secure enclosure of the laboratory has shifted elsewhere, for example, to the perimeter of buildings or campuses? Or is securitization today no longer resolved by physical walls?

The Architecture of Novartis

Daniel Vasella, a trained medical doctor, created Swiss pharmaceutical company Novartis in 1996 through the merger of Ciba-Geigy and Sandoz. Vasella stepped down as the company's CEO in 2010 and retired as chairman of the board of directors in 2013. His legacy is significant. When he retired, Novartis was ranked number one in pharmaceutical sales in the world and fourth out of twenty on the 2014 Access to Medicine Index. This independent index ranks pharmaceutical companies on how readily they make their products available to the world's poor.[10] The NIBR employ about 6,000 scientists, physicians, and business professionals across sites in Cambridge and Basel in Europe; East Hanover, New Jersey; Emeryville, California; Cambridge, Massachusetts; and La Jolla, California in North America; and Shanghai and Singapore in Asia. Research at Novartis is focused on developing therapeutics in the areas of immune-oncology, oncology, neuroscience, cardiovascular and metabolic diseases, auto-immunity, transplantation and inflammatory diseases, musculoskeletal diseases, eye care, and respiratory diseases.

Pharmaceutical companies such as Novartis operate in a very volatile environment, largely because patents on the drugs they develop expire (usually after five or fourteen years in the United States), and the development of new drugs is a lengthy, uncertain and expensive process. The average cost to a brand-name company of discovering, testing, and obtaining regulatory approval for a new drug with a new chemical entity was estimated to have increased from USD 800 million in 2003 to USD 2.6 billion in 2014.[11] The cost of development contributes to the overall cost of pharmaceuticals. However, pricing is more closely related to the degree to which a drug is perceived to improve health outcomes than to the cost of research and development or manufacture. The more painful or threatening the illness, the higher the drug's cost. Hence, the disproportionate investment in illnesses that affect the First World and its aging demographic. Once the patent has expired, generic brands move into the market to produce lower-cost alternatives. These brands are highly competitive for they have only to recoup the cost of manufacturing (and, indeed, are often manufactured by the original patent owner under a different name). In 2016, Novartis's blockbuster cancer drug, Gleevec, lost its exclusivity. Novartis had hoped that a recently launched drug might replace the anticipated lost revenue, but it failed to be taken up by the medical profession. Novartis shares lost 23 percent of their value between May 2015 and May 2016, and the company announced a restructure.[12] Many large companies took similar hits between 2012 and 2016, with seven of the top leading prescription drugs in the United States losing their patent protection in 2016 alone, an event called a "patent cliff," which affects global markets beyond the pharmaceutical industry. In the first quarter of 2018, however, Novartis's profits jumped 12 percent as it rolled out a new heart-failure drug, Entresto.[13] In this rollercoaster environment, mergers, acquisitions, and restructures occur with a rapidity that contrasts the long and slow process of research and development. In the short few years since Vasella's departure, Novartis has acquired CoStim Pharmaceuticals, GlaxoSmithKline's cancer drug business (for USD 16 billion), Spinifex Pharmaceuticals, Admune Therapeutics, Selexys Pharmaceuticals,

Encore Vision, Ziarco Group, Advanced Accelerator Applications (for USD 3.9billion), Avexis (USD 8.7 billion), and Endocyte (USD 2.1 billion).

Vasella set out to steer the company through its business challenges and to create a reassuring and palpable sense of stability and longevity. This he accomplished partly through amassing a museum-quality collection of art and antiquities ranging from African sculptures and Navajo rugs to contemporary works of art by Richard Serra and Jenny Holzer. He also collected architect-designed buildings, arguing that a good environment would make a positive contribution to the working lives of Novartis employees. He was a visionary among pharmaceutical leaders in recognizing the value of investing in research laboratory architecture.[14] While there was some concern among shareholders that the company was being financially imprudent, the investment in architecture is of a minor order given the company's scale and profits.

In Basel, Vasella set out to transform 50 acres (20 hectares) of industrial landscape into an attractive campus housing the NIBR and its corporate and administration headquarters. In 2000, Vasella employed Vittorio Magnano Lampugnani to develop a comprehensive masterplan.[15] Alongside his thirteen-year role on the steering committee of all Novartis campus sites, Lampugnani was professor of the History of Urban Design at the Swiss Federal Institute of Technology, Zurich. When he began on the Novartis project, Lampugnani was dean of the Department of Architecture at ETH and undertaking the curation of an exhibition at the Museum of Modern Art in New York on Mies van der Rohe's early career in Berlin. He had spent a decade working as the deputy editor, then editor, of *Domus* and authored several text books on twentieth-century architecture and planning. Not surprisingly, Lampugnani's experience and approach are essentially academic. His architectural practice with Marlene Dorrie had completed just one office building in Berlin and a small housing project in Graz. Steeped in European history, he is inclined to rationalist abstraction. Underlying the masterplan for Novartis is his belief in the enduring value of orthogonality and axial planning, as well as the conviction that architecture should be subservient to a larger urban project. In a 2002 lecture on the architecture of museums, a type he proposes to be symptomatic of architectural trends generally, Lampugnani decried the situation wherein contemporary museum buildings have become "pure materializations of each of their author's architectural attitudes" and narcissistic "self-portraits" indifferent to their contents and practical needs.[16] In disagreement with the pluralism he saw around him and wishing to keep check on such tendencies in the architects to be commissioned for Novartis, Lampugnani imposed a regular grid, a consistent height, and material uniformity on all future developments. It is an idealized and formal vision, promising a return to a historical urban model that, in fact, has never, existed.

Novartis describes its Basel campus as "a state-of-the-art workspace" that fosters "a dynamic environment that thrives on communication, creativity and interdisciplinary collaboration."[17] In effect, the campus is a kind of social experiment in creating what Vasella calls "an ideal atmosphere."[18] While intended to be lively and with a substantial population of 5,500 employees, the social character of the campus is, nevertheless, sterile. It is a rigidly controlled and gated enclave outside of Basel's town center. Vasella has explained that, having experienced personal attacks by "animal-rights extremists" (in 2009 the ashes of Vasella's mother were disinterred and the grave violated, and a month later his holiday home was firebombed by the Animal Liberation Front), he wanted the secure perimeter in order to allow him and his staff "to feel completely free of controls."[19] The company's nurseries for the children of employees are not on campus, and all its occupants are adults of working age employed by Novartis or official visitors. The public can visit the site only by participating in one of the paid weekend tours organized by Basel Tourismus.[20] No photography is allowed on these tours, and every participant must supply his or her address, phone number, and a photocopy of a valid identity card or passport. So, while the campus has the population, and vague structure, of a small town, during the working week it feels as though the legendary Pied Piper has passed through, taking with him children and the elderly, tourists and the unemployed.

On the positive side, the Novartis campus avoids the parking-dominated planning of many office complexes. Its very large car park is below ground. The absence of cars allows Lampugnani to recreate a pedestrian mall, complete with ice-cream stand. This is Fabrikstraße, literally "Factory Street" (figure 5.3). While the campus is made of real buildings and trees, it feels oddly scenographic and surreal, as if it were a set from Jacques Tati's film *Playtime* (1967). It does not, as urban streets usually do, foster journeys from one place to another. Fabrikstraße begins inside the entrance gate and ends abruptly at a high blank wall on the site's opposite boundary, which happens to be the border between France and Switzerland. Vasella paid USD 98 million (CHF 100 million) to the French to have them move the border crossing as it interfered with the architectural masterplan.[21]

Within Lampugnani's preexisting structure and its constraints, outstanding architects have been commissioned for individual building projects, including Kazuyo Sejima and Ryue Nishizawa of SANAA, Rafael Moneo, Diener and Diener, Peter Markli, Tadao Ando, Fumihiko Maki, Frank Gehry, and David Chipperfield. When we visited in 2014, a building by Herzog and de Meuron was under construction, and another by OMA was in the planning stages. Nikolai Ouroussoff, writing for the *New York Times* in 2009, reported that Vasella had mimicked "a formula that has become the norm

Figure 5.3
Fabrikstraße, the main street of the Novartis campus in Basel. Masterplan: Vittorio Magnano Lampugnani. Photograph by BradP, Flickr, Creative Commons.

for big-money development in cities as disparate as Las Vegas and Abu Dhabi, he has hired an army of world-renowned architects."[22] The result is not, however, a collection of recognizable signature works, as can be seen at the Vitra campus, less than four miles away. The "straitjacket" of Lampugnani's urban plan has proven an effective dampener on individual expression. Ouroussoff notes that the masterplan has produced "a row of nearly identical office buildings" and finds the project reminiscent of, indeed even more clinical than, Mussolini's urban fantasy EUR on the outskirts of Rome.[23] Paolo Fumagalli is more polite in his commissioned essay for Novartis's publication on Fabrikstraße 22. "The importance and rigour of (Lampugnani's) plan are so relentless toward exception that each architect, albeit with their own chosen design and material, has created an architecture of discipline and reflection, always pondered and controlled."[24] Frank Gehry's formally exuberant office building for the Novartis Human Resources Department is the exception. The overall impression of the campus is one of conformity and refinement, rather than architectural experimentation or innovation. Each of the buildings uses astonishingly expensive materials and detailing. Built-in furniture and lighting are customized, and everywhere there is evidence of the art collection amassed by Vasella. For example, the visitors' center, designed by Swiss architect Peter Märkli, has an integrated LED façade by American artist Jenny Holzer. The walkway from the reception building consists of forty-three glass panels with blurred silkscreened text by Austrian artist Eva Schlegel, in collaboration with the Italian architect Marco Serra.

Chipperfield's Design

The selection of Chipperfield to design the fourth new laboratory building on the campus since 2008 is fitting, given the objectives of the master plan.[25] Where other architects may have felt their wings had been clipped, Chipperfield favors a tempered neo-modernism of orthogonal lines and direct structural expression. He employs a narrow palette of materials, such as concrete and stone, to convey permanence (figure 5.4). Chipperfield worked for Sir Norman Foster as a young graduate, and his major influences are Louis Kahn, whom he often quotes, and Mies van der Rohe. When awarding him the Royal Gold Medal in 2011, Ruth Reed, then President of the Royal Institute

Figure 5.4
Exterior of Fabrikstraße 22 (2010), Basel. Architect: David Chipperfield. Photograph by authors.

of British Architects, extolled Chipperfield's architecture as one of "calm rational elegance."[26] The Fabrikstaße 22 project was designed and led by Chipperfield's Berlin office in association with Swiss architects and veterans in the planning of laboratories, Burckhardt and Partner.

Novartis requested an open laboratory clear of obstacles.[27] The Chipperfield design sets up the conditions for the client's spatial experiment by employing load-bearing façades and pre-stressed beams spanning 88 feet (27 meters) between the cores to establish generous column-free floors. In keeping with the urban aspirations of the campus, the ground floor of the five-story building is occupied by a Thai restaurant. Cha Cha Restaurant has a lofty battened timber ceiling, sturdy pale beech furniture, paneled walls, and a darker timber floor. The restaurant is furnished with long timber tables to accommodate 192 people. The bench seats alongside each table seat up to sixteen diners—a sharing experience intended to emphasize the group over the individual. The uppermost floor of administrative offices wraps around an open courtyard containing Serge Spitzer's installation *Molecular (BASEL)* (2002–2008). The courtyard is essentially a world apart, open only to the sky—a vestigial memory of the Salk Institute's plaza. Planting is a monoculture of Japanese Zelkova trees. The floor and primary material of the "roof garden" consists of fifty tons of small, green-glass spheres, about twice the size of a regular marble. Bernhard Fibicher observes, "There are no regular structures, nor any comprehensible organization of discernible system. The spheres simply form an aggregate subject to a permanent process of self-organization. Probably a pattern will never emerge, yet the sphere field is constantly moving."[28] It's an image of bubbling foam or froth that we will return to.

The first, second, and third floors house 140 laboratory workplaces in open spaces finished with polished concrete off-form ceilings and pale laminate floors. Only the circulation core of escape stairs, lifts, and toilets is enclosed in opaque walls. Laboratory benches occupy the same undivided space as the write-up desks on which sit potted plants, personal computers, and office paraphernalia. Circulation and costly equipment share the floor. Chipperfield describes this radical elimination of the division between laboratory and offices in social terms: "Avoiding closed spatial cells as far as possible and creating an open spatial situation allows all members of all working groups to communicate."[29] Novartis claims this unbounded laboratory represents "a step forward in laboratory technology development—the floors are open and without supports, with no opaque or glass separating walls."[30] Floor layouts "minimize the use of glass walls and other enclosures."[31] We have seen elsewhere the belief that an "open spatial situation" leads to openness of communication between scientists, but nowhere else has this ambition led to the removal of walls enclosing a PC2 laboratory.[32] Indeed, elsewhere we

Figure 5.5
The circulation corridor between the wet laboratories and the write-up and office spaces at the
Faculty of Science Building (2015) at the University of Technology, Sydney (2015), designed by
Durbach Block Jaggers and BVN, has black surfaces to dramatize the gap between the two. Photo-
graph by authors.

have seen architects exaggerate the differences between wet and dry laboratories, for
example, at the Faculty of Science Building for the University of Technology, Sydney
(2015) designed by Durbach Block Jaggers and BVN (figure 5.5).

Taking that social ambition to its end at Fabrikstraße 22 was a risky strategy. Essentially separation is achieved through the management of air intake and exhaust. It is
tempting to see in this project the mainstream adoption of a more radical concept of
architecture as air or atmosphere, one that dates to artist Yves Klein's proposals during
the 1950s. It would be possible to confuse this project as akin to the projects celebrated
in the special issue of *Architectural Design* called "Meteorological Architecture" (2009),
including those of the Swiss-French architect Philippe Rahm, who pursues the modulation of environments through radiation, convection, evaporation, pressure, temperature, and light. Rahm's design for the 173-acre (70-hectare) Jade Eco Park (2016) in
Bangkok is the first project to see the atmospheric ambitions of the current generation
of meteorological architects realized on a grand scale. It would, however, be a spuri-
ous alignment. The goal of Chipperfield and his clients was not to explore air as an

architectural material, nor to elicit a physiological response from those in the building. It is of little, if any, consequence that Fabrikstraße 22 deploys the same body of knowledge about the conditioning and movement of air as Rahm and his peers do. His client's ambitions, Chipperfield understood, were to eliminate the physical divisions between the wet laboratory, the equipment, and the activity of writing. Air-conditioning was the means to an end, but this does not make the approach any less daring.

Certification from the authorities was at first refused, and it required muscular persistence to secure provisional certification of six months. We imagine that the diplomatic skills deployed to move the border gate between Switzerland and France came in handy in removing the border between the laboratory and the social spaces at Fabrikstraße 22. Rigorous testing for contaminants carried out at the end of the trial period yielded positive results and the laboratory was granted certification for meeting the containment demands and fire safety regulations of the Swiss regulating authorities. For Novartis, removal of the physical barriers around the laboratory space was also met with resistance by some of their researchers. As the organization is large, it was agreed that only those scientists supportive of the new direction would work in the new laboratory building with its unfamiliar arrangement. Enclosed, nonspill cups were designed and supplied to allow scientists to continue to drink coffee while at their desks—the final hurdle. Funds were set aside to install partition walls if certification was not forthcoming or containment could not be realized. The goal of increased sociability and movement flows was considered worth pursuing in spite of the difficulties it presented.

Bursting the Laboratory Bubble

We might think of the contemporary research building as a kind of foam, a concept Peter Sloterdijk uses to explore the spatial and ontological relationships between people and their contexts.[33] In the research building, the laboratory proper and adjacent spaces for equipment, write-up, and circulation are conventionally distinct bubbles in the organization of the research building. Each has its own specifications for air quality and protocols. The envelopes that circumscribe them are shared, facing the interiors of adjacent bubbles. Information and artifacts pass between them according to a strict logic and prescribed flows. Sloterdijk calls bubbles an "inclusive exclusivity" because, much like the spaces of laboratories, each has its own internal logic but also a resistance to that which is outside.[34] Bursting the wall of one bubble does not lead simply to one less bubble while the rest of the foam remains the same; rather it sends shudders through the entire structure, which leads to its reorganization. At Fabrikstraße 22 the consequence of bursting the bubble that previously enclosed the wet laboratory space is evident in the internal stair encased in a glass tube that rises through the building. The original design proposed an open stair. Open stairs are a familiar device in biosciences research buildings, deployed for socialization. Indeed, in the laboratory for Novartis in Cambridge, Massachussets (2016), the architect Toshiko Mori proclaims its open stair provides "vibrant spaces for chance encounters."[35] When Novartis insisted on opening up the laboratory, as Chipperfield explains, the stair was now in a situation where it "connects floors with a high potential for risk" and had to be "enclosed with EI 60 fire protection glass."[36] This effectively removes the vertical circulation through the building from the work spaces—a result somewhat at odds with the desire to maximize ease of movement and connection between scientists.[37] To compensate for its enclosure Chipperfield commissioned industrial designer, Ross Lovegrove, to design a sculptural stair (figure 5.6). Lovegrove, known for his design of a spiral staircase named "DNA" (2005), fancifully describes himself as "an evolutionary biologist, more than a designer."[38] He is not; he's an industrial designer inspired by natural forms. The treads of the Fabrikstraße 22 stairs are biomorphic and curvaceous in an otherwise orthogonal and severe building. The radical openness of the combined laboratory and write-up floors also presented a challenge for intimate conversations and meetings. The solution takes the form of a glazed circular room just large enough for four people to have a private conversation that is inaudible to others (figure 5.7). Both the stair and office enclosures are achieved using a fire-protective glass partition system, designed and manufactured by Swiss company BlessArt. Used across the campus, this system produces a distinct family of bubbles with their own logic and interchangeability.

At Fabrikstraße 22, large floor-standing machines, such as microbiological incubators, environmental chambers, cryogenic storage freezers, and sample storage refrigerators, are huddled together at one end of the laboratory floor. Their arrangement is completely independent of the experiments taking place at the laboratory benches— they are not, as is more typical, grouped in enclosed rooms adjacent to relevant benches. Their self-sufficiency is evident. There is also much that is not evident. Here, scientists deploy a range of remote mobile technologies, which they use to operate laboratory equipment at a distance. They are not entirely liberated from the laboratory bench and the desk by this equipment, but are tied to it by a very much longer leash than those with desktop computers in fixed write-up spaces adjacent to enclosed laboratories.

The perimeter envelope of the building is also treated unconventionally. Floor-to-ceiling, frameless glazing spans the openings between the structural columns. The glass is recessed to align with the inside face of the columns. Just behind the glass, full-length sheer white drapes are hung on concealed tracks (figure 5.8). The curtains are not intended for privacy since the building is already buried deep in the exclusive

Figure 5.6
The whalebone stair designed by Ross Lovegrove in its glazed tube, Fabrikstraße 22 (2010), Basel.
Architect: David Chipperfield. Photograph by Anton Tutter.

Figure 5.7
Glazed meeting rooms create a "cone of silence" for scientists in the otherwise open space at Fabrikstraße 22 (2010), Basel. Architect: David Chipperfield. Photograph by authors.

campus. Nor do the drapes provide sun protection or thermal insulation. Fabric of any kind is unusual in a program whose interior design is conventionally directed at hygienic efficiency. Decorative drapes of this length and translucency are more typical of luxurious hotels and important civic buildings—Chipperfield often uses them for these building types. There is more going on here than achieving sensual beauty or connotations of wealth or leisure. Chipperfield describes the continuous curtains as "an experimental arrangement." It would not be unreasonable, given the architect's reputation for sober precision, to consider this seemingly trivial interior detail to be integral to the social and spatial experiment at the heart of this project.[39] It is not an afterthought introduced by an independent interior decorator; indeed, it is the only sensual and soft moment in an otherwise severe interior landscape. What the curtains do, their primary purpose, is to obscure and blur the actual limits of the space. The curtains simultaneously erase and multiply the surface or limit; they have a three-dimensional form but no definable surface. They dissolve into multiple threads with infinite microsurfaces. The edge of the building is barely perceptible in this softly glowing, white cocoon. There is a sense of floating in an unknown, potentially unlimited

Figure 5.8
The curtained perimeter of Fabrikstraße 22 (2010), Basel. Architect: David Chipperfield. Photograph by authors.

space, reminiscent of the "white limbo" spaces devised by George Lucas for his first commercial film, *THX 1138* (1971), which recalls Zamyatin's *We*. In this way, the dissolution of the boundary of the laboratory bench space is extended farther and the laboratory's limits perceptually deferred.

Simulated Science

Gilbert Simondon has taught us to scrutinize the small material changes in technical apparatus over time. In his discussion of the processes of technological development Simondon offers this example: in the development of electronic tubes, the passive components become reduced, whereas its active parts are condensed. The rubber base grows smaller and eventually disappears. At the same time, the functional parts of the tube start to take over more space within the glass ampoule. In regard to such objects, Simondon speaks explicitly about their "morphological evolution" using visual strategies developed in neo-Darwinism (although not by Darwin himself). He calls this process "concretization," whereby the parts of a technological object develop independently and iteratively. He does not mean that objects become stable and reach a final form, but that objects become increasingly self-sufficient and internally coherent as their internal parts change. Simondon writes:

> A concretized object … becomes independent of the laboratory with which it is initially associated and incorporates it into itself dynamically in the performance of its functions. Its relationship with other objects, whether technical or natural, becomes the influence which regulates it and which makes it possible for the conditions of functioning to be self-sustaining. The object is, then, no longer isolated; either it becomes associated with other objects or is self-sufficient, whereas at the beginning it was isolated and heteronomous.[40]

For Simondon, the laboratory is a technological ensemble consisting of a variety of machines and devices that need to be isolated from one another to work properly. It is a regulatory environment or exterior for technical objects. Since Simondon first wrote of the relationship between technical objects and the laboratory, a new genre of technical objects has entered the ensemble and affect communications between diverse machines, as well as machines and scientists. The automation of some of the repetitive tasks in the laboratory using robotic arms is widespread, but there is more going on than laboratory robotics. The emergence and proliferation of information and communication machines have changed the relationship between machines and the laboratory and spawned concepts such as the virtual laboratory, "collaboratory research," "citizen science," and e-science. E-science tools such as shared databases are not stand-alone technologies, but are tools developed to be used in networked mode and accessed in different locations.

The idea of e-science is inseparable from the idea of the "collaboratory." Defined by William Wulf in 1989, e-science is a "center without walls, in which the nation's researchers can perform their research without regard to physical location, interacting with colleagues, accessing instrumentation, sharing data and computational resources, [and] accessing information in digital libraries."[41] Wulf's definition has been much repeated in different forms, often simplified as the concept of the "laboratory without walls." Collaboratory advocates envision the demise of physical laboratories and a revolutionary shift in scientific practices. For example, in a *Science* editorial Richard Zare wrote that science is entering an era when "knowledge is available to anyone, located anywhere, at any time; and in which power, information and control are moving from centralized systems to individuals."[42] Detractors point to the ways in which scientific social hierarchies, competition, intellectual property concerns, the ease of face-to-face communication, and the importance of visual cues in conversations all mitigate against the vision of an open virtual scientific space. Clearly in some fields, such as genetics, there has been increasing reliance on, and benefit from, large-scale peer-to-peer internet-accessible DNA databases, such as the nonprofit Global Alliance for Genomics and Health. Information and communication technologies have become central to genetic research, while in other fields there is less uptake and knowledge sharing. But even in genetics there are naysayers who are reluctant to share genetic data, citing privacy concerns for patients.

Virtual laboratories are seen as a means for diversifying collaboration beyond the limitations of organizational, disciplinary, or national boundaries. They are a way of remotely accessing limited research tools and equipment to perform or participate in experiments, as well as to store and share information. There is another aspect to the virtual laboratory. The simulation of research subjects—virtual plants, animals, and humans, as well as parts of these such as organs, molecules, and cells—is an emerging technology that combines biological knowledge, mathematical formulations, and computer graphic techniques. Complex systems can be studied by developing models of their underlying characteristics on a computer and using computationally intensive methods to learn about and predict the behavior of those systems. These methods are called "simulations" or "numerical experiments." Debates about the relationship between scientific theory, simulation, and physical testing have been raging, inconclusively, for the past two decades. Some take the view that simulation is a brand new "third mode" of science, neither experimental nor theoretical.[43] Simulations are based on models, drawn simultaneously from theory and the physical world. They exceed previous scientific models or diagrams to produce models of phenomena that look much like those made in laboratories—for example, images that resemble photographs of material physical models. Simulations are constructed from existing data but are capable of producing new data for analysis. At present it is possible to design computer models that recreate the structure and simulate the development and physiology of plants, not only in isolation, but also in their ecological contexts.[44] In the biosciences, rapid accumulation of biological data from genome, proteome, transcriptome, and metabolome projects has resulted in sufficient information about a living cell's molecular components to develop software platforms for whole-cell-scale modeling, simulation, and analysis.[45] As early as 2001 researchers in the biosciences were declaring that "it is no longer purely speculative to discuss how to construct virtual cells in silico."[46] In 2006, Gireg Desmeulles and colleagues described a new type of investigation for predictive complex biological systems, which they call the "in virtuo" experiment.[47]

The advent of the "in virtuo" may yet herald the end of controversial practices such as vivisection, but we would be wise not to exaggerate the difference between the virtual dry laboratory and the wet laboratory. As Knorr Cetina observes, laboratories have never dealt with natural objects as they are, but with objects detached and installed in a new environment where they are manipulated and reconfigured on terms determined by the scientific experiment.[48] In a sense, science has always produced artificial, mediated, if not virtual subjects that are made workable in their relationship with scientists. In the laboratory, scientists construct "object images or their visual, auditory, electrical, etc., traces, with their components, their extractions, their purified versions."[49] Objects in the laboratory become representations of real-world phenomena or processed partial versions of these phenomena. Transcribing these into text, image, and model makes the subjects of experimentation continually present and available for inquiry. Digitization extends the availability of this same data to the whole of the scientific community. In any case, since the publication of Norbert Wiener's *Cybernetics* (1949) and the discovery of DNA, molecular biology has viewed "the bodies of humans and other animals as information systems, as networks of communication and control."[50] In this context, discovery or innovation might be an effect, not so much of bringing scientists together in physical settings, but of eliminating the differences between the organic and the machinic.

The company that eliminates the differences between the organic and machinic, thereby transforming employees and consumers into cyborgs, is likely to be the most successful in producing new knowledge. Daniel Vasella knew this when he left Novartis to join XBiotech, a Texas company developing therapeutic antibodies cloned from donors who possess natural immunity against certain targeted diseases. Novartis is aware of this too. Vas Narasimhan became the CEO of Novartis in February 2018. Just before he took up the role, Narasimhan proposed that the company would cut between

10 and 25 percent of the cost of trials through effective use of digital technology. He would do this by acquiring or partnering with artificial intelligence and data analytic companies.[51] "I really think of our future as a medicines and data science company, centred on innovation and access," said Narashimhan in a 2017 interview.[52] "I believe immensely in data and digital," he said in another.[53] The company's expertise will shift—2200 employees in Switzerland will lose their jobs by 2022, replaced by 450 new jobs in the field of personalized medicine. Its relationship to consumers and patients will change, too. Novartis will use "advanced analytics" to recruit "patients" for clinical trials and monitor trials in real time using predictive algorithms. From primarily manufacturing pills, Novartis plans to move increasingly toward personalized gene therapies.

A decade after Fabrikstraße 22 was conceived, it can be seen as a harbinger of what was to come for the business and for experimental science. The continuity of its laboratory benches and write-up areas asserts that there is no longer a gap between the physical and the computational. Science today operates with digitized matter and materialized information. Fabrikstraße 22 registers the loosening of science's debt to the functional space of the laboratory as a secure site for physical experimentation. The conventional linear direction from action in the laboratory to recording in the write-up has been replaced by a space of multidirectional vectors. The laboratory has always existed alongside and inside multiple sites of observation, collection and manipulation—from Darwin's *Beagle* to Craig Venter's sloop, *Sorcerer II*. But the *ideal* of the laboratory as a physical and bounded place has persisted, until now. At Fabrikstraße 22 we see the laboratory, not so much disappear, as expand to the point where the laboratory is now everywhere: in virtual models and algorithms, in patients and scientists. In the enormous interior of the Crystal Palace, Sloterdijk recognized "nothing less was at stake than the complete absorption of the outer world into an inner space."[54] At Fabrikstraße 22, it is the inverse to which we now bear witness. The Crystal Palace conjures for Sloterdijk the "idea of a building that would be spacious enough in order, perhaps, never to have to leave again."[55] Fabrikstraße 22 gives architectural figuration to the idea of a laboratory unbounded, a ceaseless experiment that knows no bounds.

Part II: Expression

The relation between architecture and science has undergone a radical transformation in recent decades. While architecture continues to house the "content" and work of the biosciences, its ambitions are now focused on education, socialization, articulation, and all matters of "expression." The boundaries that were once conceived of as accommodating or containing science come to operate as placards, telling stories of the ideas and ideals of science, sometimes simplistic and sometimes complex. These expressions are directed toward the public, philanthropists, and funders, and sometimes are reflected inwardly to the scientists themselves. Scientific imagery and analogies become the playthings of architects, fighting to find appropriate ways of representing the often microscopic complexities of science.

The following four chapters explore the rhetorical function of the contemporary laboratory—how the buildings speak and what they speak of. In doing this, we go beyond surface treatments and the articulation of boundaries to include the organizational, formal, tectonic, and atmospheric qualities of the laboratory. We trace the use of analogy and metaphor—of images of virus cells upscaled into floating meeting rooms, of DNA strands that become stairs and printed images applied to glazing. We consider analogies that operate in complex manners—metaphors of crystals and trees that are more romantic than real, and on occasions inverted. Such expressions are not as simple as they might seem and enact a reciprocity between lay ideas about nature and scientific understandings of life. Thus, the expression of the biosciences in laboratory architecture is in many ways a political problem—a problem of what can be said to whom, and indeed what shouldn't be pronounced or made visible. We consider the proliferation of metaphoric animals in architecture and the real laboratory animal as a case in point. We also consider the impossibility of tying the content of the biosciences to architectural expression and the manner by which architects might negotiate this impasse.

6   Enunciation

The sight of scientists carrying out experiments behind glass walls in the laboratory is, it seems, frequently found rhetorically lacking by architects and their clients. An additional layer of representation is called for, one that attempts to translate scientific purpose and practice through the selection and application of scientific imagery and analogy. Architect Will Alsop captures this desire when describing the Blizard Building (2005), a new research center for Barts and the London School of Medicine and Dentistry, at Queen Mary University of London. Here, Alsop declares, "The very fabric of the building speaks about science" (or, as we will argue, *for* science).[1] Barts and the London School of Medicine and Dentistry was formed from two of the UK's oldest medical institutions, St. Bartholomew's Hospital Medical College and the London Hospital Medical School. The Blizard Building was named after the founder of the London Hospital School, Sir William Blizard. Though named after a historical figure, the Blizard Building speaks of the contemporary state of science. Within its double-height glazed pavilion wing is a helical stair and four suspended pods that emulate different cells and molecules—named Centre of the Cell, Spikey, Mushroom, and Cloud (figure 6.1). The mews between the building houses a recently completed fifth pod named Neuron, and the glazed curtain wall of the Blizard Building is adorned with large abstract images appropriated from molecular science by artist Bruce McLean. It is a showcase of analogical delights.

The Blizard Building, while one of the most graphic examples of the genre of analogically expressive laboratories, is not exceptional. The colorful rows and columns of genome sequencing are referenced in the pattern of glass and metal panels on the façade of the Patriarche & Company–designed Genzyme Labs (2010) in Lyon, France. In the redevelopment of the Walter and Eliza Hall Institute of Medical Research (WEHI) (2012) in Melbourne, Australia, architects Denton Corker Marshall added a perforated metal façade with an abstract pattern of holes representing DNA and an animation

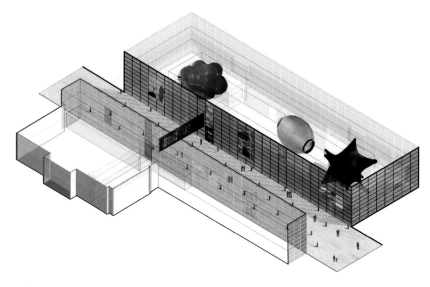

Figure 6.1
Volumetric diagram of the Blizard Building (2005). Architect: Will Alsop. Drawings by research team.

wall along the main entrance displaying the work of Drew Berry. Berry's visualizations for WEHI model the properties and behavior of DNA strands. Images of molecules and cells ornament the Robert Ho Research Centre (2011) in Vancouver, Canada, by CEI Architecture and MCMP Architects. The Hawkins/Brown designed biochemistry building for Oxford University includes inkblot images in its façade by artist Nicky Hirst, which recall the paired shapes of human chromosome mappings. In emulation of the X chromosome, Lyons Architects included two stairs crossing each other in the Biosciences Research Building (2012) at the Australian National University in Canberra. The building also incorporates a tessellated façade based on the hexagonal geometry of a molecular structure, a motif Lyons Architects repeat a year later at the La Trobe University Institute for Molecular Science (2013) in Melbourne, Australia. A giant mural is featured on the east façade of the Anna Spiegel Research Centre for Molecular Medicine Building (2011) in Vienna by Kopper Architektur. Here, the artist Peter Kogler contributed an enlarged and abstracted vision of an infinitely extensive cellular organic world. The façade of the Science Laboratories (2010) at the Chinese University in Hong Kong by RMJM portrays a giant version of the periodic table of elements. Renzo Piano's Jerome L. Greene Science Center (2016) for Columbia University in New York features along its main corridor a digital interactive art installation that animates the work

Figure 6.2
International Neuroscience Institute (2000), Hanover, Germany. Architect: SIAT Architekten +
Ingenieure. Photograph by Christian Schd, Wikipedia Creative Commons.

going on elsewhere in the building for visitors and scientists alike. Such examples treat
the façade as articulate surfaces, like pages in a textbook that communicate to others.
There is a host of buildings that speak in three dimensions, but still defer to analogy and metaphor to do so. Such buildings include the two brain-shaped neuroscience research and treatment buildings conceived by neurosurgeon Majid Samii, one in
Hanover, Germany (2000) (figure 6.2), the other in Tehran, Iran (2019); the Cornell
Ornithology Laboratory (2009) by RMJM that takes the form of an abstract bird; and
the cell-shaped floor plan of the DZNE German Center for Neurodegenerative Diseases
(2017) in Bonn, Germany by Wulf Architekten. We could go on—there are many more
examples. The myriad ways in which scientific images and artefacts are appropriated
for architecture speak both to the potency of the images and the creativity of artists and
architects. But what do these elements communicate and how do they work? Before we
can make sense of the Blizard Building, we need to first examine two moves: the transformation of representation into modulation; and the transformation of the biological
into the architectural.

From Representation to Modulation

The discourse of boundaries and their concomitant representational agenda are described by Gilles Deleuze in his "Postscript on the Societies of Control" (1990).[2] Here, the divisions Michel Foucault identified between disciplines and between those institutions that regulate human endeavor (educational, health, medical) through architecture (the school, the clinic, the hospital) become ever more fine-grained. In his earlier text *The Logic of Sense* (1969), Deleuze noted the serration that once marked boundaries had become the element of an articulation.[3] It is a theme he advances in his "Postscript." One can imagine Deleuze describing the now-seamless articulations of the contemporary biomedical laboratory when he refers to the operational modes of the control society:

> The different internments or spaces of enclosure through which the individual passes are independent variables: each time one is supposed to start from zero, and although a common language for all these places exists, it is *analogical*. On the other hand, the different control mechanisms are inseparable variations, forming a system of variable geometry the language of which is *numerical* (which does not necessarily mean binary). Enclosures are *molds*, distinct castings, but controls are a *modulation*, like a self-deforming cast that will continuously change from one moment to another, or like a sieve whose mesh will transmute from point to point.[4]

Deleuze's target is the manner by which all of society's institutions continue their hold over us even after we have technically left them. Facebook keeps us in contact with our schools even as we operate in professions. Universities perpetuate the dogma of lifelong learning via their alumni-linked Twitter accounts. Our mobile communication devices keep us working well after the hours for which we are paid. Those same devices keep our lovers a mere gushing emoji away during tedious meetings with administrators. The disciplinary society Foucault had described relied heavily upon demarcation and fixation (molds). The control society, on the other hand, relies on translation, transmutation, and transformation (modulation). For Deleuze, the relation between all the independent variables we move through, in a day or across a lifetime, are *anological*. The analogical is best thought of in terms of a logic that brings discordant elements together. That is, the relation between parts (*ana*) is determined in word or speech (*logos*). Analogy involves a simple operation. It is a logic that suggests that if element x is the same as element y in one respect, then element x is likely to be similar to element y in others. Analogy generates a consistency to the otherwise inconsistent. It modulates the molds.

Much of the architecture of the biomedical laboratory relies heavily on the function of analogy to make ideas, elements, and practices that are discordant appear consistent. The architecture posits itself as a sliding scale from the microscopic to the human and to the community. The DNA stair, which is repeated *ad nauseum* in laboratory buildings around the world, is a case in point.[5] No stair is a strand of DNA, and there is no shared functional relation between the stair and its microscopic equivalent. Nevertheless, the analogy works to transform bodies as they pass between the building's successive levels, on steps labelled A, T, C, and G. Just as a microscope slides molecules into visibility, an architect slides such images onto the patterned façade of a building or the form of its components. So, too, are bodies slid along DNA stairs and transformed in the process. Through analogy the spectator becomes the scientist, the visitor a patient, the citizen a scholar, and vice versa. What is at stake in this process?

Biological Analogy in Architecture

It is serendipitous that Philip Steadman commenced his seminal work *The Evolution of Designs: Biological Analogy in Architecture and the Applied Arts* (1979) with a quotation taken from John Tyler Bonner's short essay "Analogies in Biology" (1963): "When I was no more than a boy and beginning to show some interest in living creatures I can remember being sternly warned by my elders to beware of the dangers of analogies. It was said in the same tone one might tell someone not to eat a certain kind of mushroom."[6] Bonner was an evolutionary biologist and expert on cellular slime mold. He is, alongside Stephen Jay Gould, famous for challenging a number of the tenets of neo-Darwinism in suggesting evolution may operate via mechanisms other than natural selection. Bonner, like Gould, had a particular interest in the formal and material mechanisms by which evolution occurred, particularly morphogenesis. Bonner's most recent book, *Randomness in Evolution* (2013), suggests his enthusiasm for the topic did not wane over the last fifty years.[7] In *Analogies in Biology*, Bonner argues for a "functional" approach toward comparative anatomy, which emphasizes the analogies between parts of different organisms (such as the wings of bats, birds, and insects), rather than focussing upon the various and wondrous manners by which flight itself can be engaged. His point relates to what is called the "sign theory" of morphogenesis, which implies that bodily form is valuable only in its ability to signify points in the tracing of the sequences of heredity.[8]

Objectors to the sign theory of morphogenesis present anatomical data as evidence to suggest that the diverse complexity of animal forms cannot be attributed to a single cause (or axiomatic theory, such as natural selection). D'Arcy Wentworth Thompson was one early objector. His biological treatise *On Growth and Form* (1917) presents the multiplicity of forces that might constitute an organism: geometrical constraints, principles and patterns of growth, mechanical forces, and so on. Darwinism turned away from the unit of the sequence of evolution to the sequence itself, and Thompson set out to counter this move.[9] Bonner and Gould share his concern. The abridged edition of *On Growth and Form* (2004) was edited and introduced by Bonner and contains a foreword by Gould. The appeal of Thompson's work to Bonner and Gould relates to Thompson's sense that "natural history deals with ephemeral and accidental, not eternal or universal things; their causes and effects thrust themselves on our curiosity and become the ultimate relations to which our contemplation extends."[10]

Many architects have admired Thompson's work. Steadman argues that of all works on biology, it is Thompson's *On Growth and Form* that has most directly stimulated architects.[11] Louis Kahn, the architect of the The Salk Institute for Biological Research, was intrigued by Thompson's analysis of bone structures. Most architects, however, are enamoured by the demonstration of transformation from one type of fish into another using a mathematical grid (without any recourse to natural selection) (figure 6.3). Thompson's drawings evoke for architects the iterative adjustments to form that often

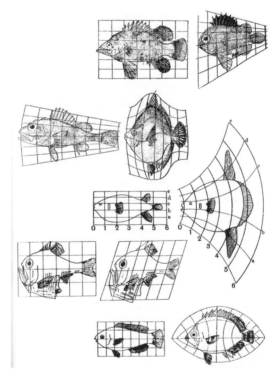

Figure 6.3
D'Arcy Wentworth Thompson, transformations of fish, from *On Growth and Form* (1917).

characterize the relations between buildings.[12] They inspired Nicholas Negroponte to declare the imminent invention of a machine that "could procreate forms."[13] For John Frazer, author of *An Evolutionary Architecture* (1995), morphological transformations are ideal for digitization, and this line of thought continues in the work of architects such as Greg Lynn. D'Arcy Thompson's drawings are, thus, oft cited as a precursor to digital parametricism. This is clearly not Thompson's point. In *Biological Analogy*, Bonner is concerned not only with single theories that swallow too many areas of investigation, but also with the dangerous tendency of analogy to overextend. He is wary of analogies that operate as a system of logic and traverse radically different scales, such as the comparison of the social structures of insects and the divisions of labor of the human animal. Such an analogy across scales is as dangerous as an axiomatic theory that collapses multiplicities under unities. Bonner suggests that "playing with analogies would seem to be a dangerous sport only to be attempted by the careful."[14]

It is not that Steadman, as an architectural historian, is careless in his engagement with analogy. His book studiously documents the history of architecture's indulgences in biological analogies. He organizes the analogies under the headings of "The Organic," "The Classificatory," "The Anatomical," "The Ecological," and "The Darwinian." Later chapters explore the way analogies have contributed to the discourse of architecture on matters such as the evolution of decoration and the theorization of design as a process of growth. Steadman introduced the complex manners by which analogy might be thought, particularly in respect to the functionalist dogma that has preoccupied both architecture and biology for the last 150 years. An example would be his explanation of Richard Owen's distinction between *analogue* and *homologue* from 1848. Analogue, Steadman relays, is "a part or organ in one animal which has the same function as another part or organ in a different animal" such as the wings of a bird and a bat, which have different evolutionary origins but perform a similar function in flight. A homologue is "the same organ in different animals under every variety of form and function," such as the four limbs of tetrapods. The four limbs of mice, bats, and crocodiles are the same organs but have far from the same form or function.[15]

This distinction between analogue and homologue bears an astonishing correspondence to the distinction between *syntagm* and *paradigm* in Ferdinand de Saussure's *Course in General Linguistics* (1916). Saussure was "concerned exclusively with three sorts of systemic relationships: that between a signifier and a signified; those between a sign and all of the other elements of its system; and those between a sign and the elements which surround it within a concrete signifying instance."[16] He stresses the point that meaning arises from differences between signifiers and that the differences are either *syntagmatic* (relating to association and positioning) or *paradigmatic* (relating to substitution). Saussure refers to syntagmatic differences as "associative relations."[17] Such distinctions, which arrived in the discourse of biology in the mid-nineteenth century and in linguistics in the early twentieth century, *should have* become important to architects, architectural critics, and theorists engaged in all form of associative and substitutional relations. But, despite notable engagements with Saussurian linguistics in the 1970s by architects such as Peter Eisenman, mainstream architecture continues to refer to all manner of substitution and appropriation of the elements of one assemblage into another as "analogy."

Outside of architecture, contemporary understandings of analogy have built broadly on Saussure's insights, but with a significant shift. The emphasis is not so much on *what* is signified as it is on the multiple associations that are brought into play and the productive value of these associations. In more recent historical and comparative works in linguistics, analogy is understood as the process of regularization that affects exceptional forms in the grammar of language.[18] Irregular forms tend to become regular, in a process that can be heard in the utterance of young children in such forms as "deads," "eated," "fighted," which are coined through association with regular plurals and past tenses.[19] Similar to the biological definition, likenesses or similarities in this interpretation are considered structural or systematic. That is, likeness and similarity come to be taken as signs of commonality. When we concentrate on similarity, it is *difference* that suffers. The similarity between the stripes of a zebra and the stripes of a zebrafish might imply similarity—and a connection between the two. However, to concentrate upon such a similarity would belie the obvious differences between an animal that roams the savannah and one that occupies the aquarium.

The structural or systematic characteristic of analogy, because it opens meaning-making to contestation, places it clearly in the realm of politics. In his 1981 text *The Political Unconscious*, Fredric Jameson refers to the structural limitations of analogical readings. He suggests, "Interpretation is not an isolated act, but takes place within a Homeric battlefield, on which a host of interpretive options are either openly or implicitly in conflict."[20] Jameson points to strategies of containment, whereby analogical readings project the illusion of authenticity and coherence. To argue by analogy is to imply that if one thing is in some respects similar to another, then the two things correspond in other, as yet unexamined, respects. We are, of course, guilty of this, as we add to the analogies already proposed by the architects of laboratory buildings, a multitude of figures; zebrafish, islands, accountancy spreadsheets, fossils, theaters, and so on. In logic, reasoning by analogy is a form of nondemonstrative argument that, unlike induction proper, draws conclusions about the nature of a single unknown thing from information about what it resembles. It is a form of reasoning particularly liable to yield false conclusions from accurate premises. This was Bonner's fear (expressed with the aid of an analogical mushroom).

In the introduction and afterword to the revised edition of *The Evolution of Designs* (1979, 2008), Steadman remarks on the changes in engagements with biological analogy that have swept through architecture since the book was first published. Steadman notes:

> The trouble with biological analogy in architecture in the past is that much of it has been of a superficial picture-book sort: "artistic" photos of the wonders of nature through a microscope, juxtaposed with buildings or the products of industrial design. But analogy at a deeper level can be a most fundamental source of understanding and of scientific insight, as many writers on that subject have pointed out. The conclusion of J. T. Bonner's essay on "Analogies in Biology," from which the terrible warning at the beginning of this chapter is drawn, is that though analogies are certainly hazardous—they are the stock-in-trade of quacks and crackpots—at the same time, if made with sufficient care, watching always for where the analogy breaks down, they can be a most fertile source of new ideas and knowledge.[21]

This is an optimistic perspective of contemporary architectural engagements with biological analogy. It is also an optimistic reading of Bonner's essay. Steadman notes "a great flowering of new theory in architecture and design, looking not just to understand and imitate natural forms, but seeking insights at deeper levels into biological processes, from which designers might derive models and methods."[22] He reflects on the impact of computer technology on the area and "important theoretical debates going on about cultural and technological evolution," which he suggests were heralded by the neo-Darwinist Richard Dawkins.[23] Steadman opines, "Modern research in 'biomimetics' (engineering analysis of organisms and their behavior with a view to applying the same principles in design) gives a new name and new rigour to what went under the banner of 'biotechnique' or 'biotechnics' in the 1920s and 1930s."[24] These new technologies are undeniable. That contemporary engagements with biological analogy are occurring under a "new name" is also clearly true. However, to claim such engagements are occurring with a "new rigour" may be another false conclusion drawn from accurate premises. This is not to say that architects should, or could, be accurate and rigorous, but that the logic of analogy itself, as Bonner warned, tends toward over-extension.

The Blizard Building

One of the key remits of the architecture of contemporary bioscience laboratories as places of massive public investment has become public engagement and information dissemination. The willful expressions of Alsop's Blizard Building involve all manner of imagery emerging out of the biosciences. The Blizard Building functions at once as a laboratory, a classroom, and as an image-generating machine—producing a series of stories about the nature of the biosciences for public consumption. One of the stated aims of the project is "to create a building which broadcasts its purpose achieved by the development of a selectively transparent building envelope.... Here the fabric of the building 'speaks' about science and creates an environment which stimulates the imagination."[25] What is of interest here is the manner by which the Blizard Building "speaks." The Blizard Building engages in transparent façades, but not so as to expose its inner workings—the laboratories are largely underground. Transparency here reveals the building's bright biological analogies. These analogies are not spoken of in subtle or gentle manners. The Blizard Building blurts things out like a child. It is a building that unabashedly speaks like children's author Dr. Seuss (and dons the colors and shapes of his illustrations) in its engagement with biological analogy.

Located in Whitechapel, London, the Blizard Building occupies a site near the Royal London Hospital and abuts the Queen Mary Innovation Centre and a Clinical Science Research Centre. A plaza, referred to by the architects as an "open mews," runs approximately eighty meters from Newark Street on the north to Walden Street on the south (figure 6.4). This mews looks much like a pedestrian street, except one is keenly aware of being between two buildings that belong together. As such, the space belongs to the Blizard, and this territorial ownership is reinforced by a recently opened Neuorn pod (2019) that stands like an engorged three-legged hedgehog over the mews, and strategically placed steel bollards. Indeed, the two buildings, along with the mews and its pod, are what is collectively referred to as "the Blizard Building." The buildings appear from the mews to be separate orthogonal multistory structures that share a common structuring grid for their façades. One façade is entirely glazed, while the other is glazed only at its ground floor, with an opaque steel screen above. The glazed areas for both are more transparent on a darker day or at night than they are in the sunshine.

Over the mews, the footbridge running between the two buildings and passing into the Neuron pod is encased in pink colored glass. Twelve bright glass panel paintings on the façades are credited to artist Bruce McLean, with whom Alsop frequently collaborated. Inscribed in a lower-case cursive within the graphic panels are the words "Genome," "Lipids," "Proteins," "Networks," "Pathways," and "Architectural Genes," which "were chosen by Professor Mike Curtis and Professor Fran Balkwill for their multiple meanings."[26] We are not sure what the multiple meanings for "lipid" might be, but we are curious and imagine that curiosity is the point. The "architectural genes" image has a black background, one shade darker than the glazing, upon which are superimposed a number of colorful amorphous blobs and two geometric shapes—a

Figure 6.4
The mews between the Blizard Building (2005) showing the Neuron pod (2019). Architect: Will Alsop and aLL Design. Photograph courtesy of Matt Brown/MattfromLondon.

triangle and right angle (figure 6.5a). It seems to be a conceptual drawing for the design of the Blizard Building itself, rather than a scientific image.

The building on the west side of the mews is narrow and tall. It is referred to as the "Wall of Plant." Just under 20 feet (6 meters) wide, this building contains a public entry, reception foyer, and a cafeteria on the ground floor. The mechanical plant services are on three upper floors. The Wall of Plant was to have a walled garden on its roof, but this remains unrealized, leaving the plant in the building's name to refer only to its hidden inner machinery. Toward the north, facing Newark Street, the slim and tall volume accommodates a four-hundred seat lecture theater and associated amenities.

The building on the east side of the mews is over 75 feet (23 meters) wide and houses the main laboratory and write-up spaces. It also holds several suspended volumes that are follies in both the architectural sense of a costly ornamental building and a kind of foolishness. A floor plate around 16 feet (5 meters) wide runs around the periphery of the volume and contains two levels of open-plan area for administrative functions and write-up. In the middle of this space is a 40-foot (12-meter) wide void, which contains

Figures 6.5a and 6.5b
The façade of the Blizard Building (2005), from both directions. Architect: Will Alsop. Photographs by authors.

four suspended forms (figure 6.6a, figure 6.6b, and figure 6.7). At ground level there is a curvy-walled, bowl-shaped pod called the "Mushroom" from which a single strand DNA stair hangs down into the laboratories below. Above, suspended in the massive void are a bulbous orange pod known as the "Centre of the Cell," a black pod called "Spikey," and a white ovoid pod named "Cloud." Each of these four objects can be occupied. The Mushroom is a space where scientists can wait for colleagues, while Spikey and Cloud are small meeting spaces accessed from the staff areas. The largest, the Centre of the Cell, encloses a classroom, referred to as an "interactive educational facility." Its purpose is to inform the public about modern medical research. Public access to the bulbous orange cell is through the Wall of Plant building, via the footbridge that runs above the mews and connects with the Neuron pod.

At the base of this theater-in-the-round is a laboratory. Benches are organized like well-lit display cases. Here the scientist has, it seems, returned to the basement, not as the maker of worlds, but as the subject of public gaze. This basement doesn't stop at the edges of the building in which we now stand, but extends under the mews and the entire site. Small circular skylights embedded in the pavement of the mews hint at this subterranean space. This massive basement accommodates both that part of the laboratory on display to the public from above and a more tightly controlled set of lab spaces that occupy a "blind-spot."

Size and Slide

There is, in the Blizard Building a disarming lack of pretension. Its analogies are not conceived within the self-conscious architectural discourse of postmodern irony or the patronizing "double-coding" of elements advocated by theorist Charles Jencks as way for architects to communicate with each other and the broader public. Spikey looks exactly like one imagines a virus cell should look—black and menacing with spikes. It's a tensile cable structure clad in a fabric that resembles the taut latex of fetish wear. The Cloud pod is shaped much like a mitochondrion cell—it is constructed from a series of connected elliptical steel hoops, which form an ellipsoid form. Externally it is clad with a white tensile fabric, while the interior is lined with oak-veneered panels in a geodesic arrangement. The Cloud is acoustically sealed for privacy but operable panels allow controlled daylight in and views out. The form of the Centre of the Cell is suggested by its title—just exponentially bigger and brighter. Its shape is based on a family of repeated components, which together form a variation on a geodesic structure. It is clad in curved composite glass reinforced plastic (GRP) panels with a glossy metallic-orange gel-coat finish.

Figures 6.6a and 6.6b
The Cloud and Spikey (opposite) above the laboratories in the Blizard Building (2005). Architect: Will Alsop. Photograph by authors.

Figure 6.7
The interior void of the laboratory wing of the Blizard Building (2005), with the Centre of the Cell hovering above the basement laboratories. Architect: Will Alsop. Photograph by Steve Cadman.

Alsop describes the pods as "a family of objects floating freely in the large, airy volume" and says that they are "intended to evoke associations with cellular and molecular forms."[27] They are described, thus, in terms that relate to Saussure's syntagm. Saussure suggests of the syntagm that "there are always larger units, composed of smaller units, with a relation of interdependence holding between both."[28] The Blizard Building forms a massive theater in which these objects are suspended. The larger structure and its function and fixation (biosciences) condition the understandings we have of the pods. Their meaning is constructed relationally. That is, the purpose and the fixations of the Blizard Building do not completely define the way in which such structures "speak."

In *Will Alsop—The Noise* (2011), architectural writer Tom Porter suggests these objects are like "fugitive abstract elements from a Roy Lichtenstein Pop Art painting from the sixties," a "menagerie of blobs and pods," and a "cavalcade of Alsopian 'street creatures'."[29] Porter goes on to describe Spikey as being "like a stranded sea urchin doubling as a giant dental care device."[30] There's an optimism here. It is the creative optimism that allows an analogy to roam beyond its context. It is an optimism that makes a homologue from the analogical. If these spaces were floating along the route of Macy's Thanksgiving parade, then maybe they would be like "street creatures." If Spikey would find itself in a marine biology lab then maybe it would be a sea urchin, just as it may be a dental device if it were found in a mouth. But it's not. Spikey is suspended in the Blizard Building and care seems appropriate in demarcating the limits to analogy.

Etymologically "analogy" comes from the Greek ἀναλογία (analogía), *ana* and *logos*, *ana* referring to an association "upon or according to" and *logos* referring to "word, speech or reckoning." *Analogos* was the word used to designate "proportion" or the "proportionate." This meaning was largely unchanged in Latin, but in late Middle English came to be associated also with appropriateness and correspondence. There is, thus, a sense of measure implied in analogy, just as there is a sense of appropriateness implied in every appropriation. When Bonner warns of the dangers of analogies it is because analogies can go too far or be taken out of proportion. The architect of the Blizard Building, however, had neither aesthetic reservations nor concerns about the dangers of representation. Indeed, Alsop once announced, "I've never seen an ugly blob."[31] This building has little sense of proportion. The void is huge, and the write-up spaces are jammed to the periphery. The cellular pods are not in proportion to each other, nor to the open-top mushroom, nor to the void itself. When one thinks about it, either the pods slide seamlessly between scales without any sense of measure or the visitor to the Blizard Building does.

The architects may be indulging in the long creative tradition of imagining what it is like to be huge or to be small—to be Alice in Wonderland, the Little Prince, or Gulliver, or to be a homunculus. This was the departure point of Bonner's book *Why Size Matters: From Bacteria to Blue Whales* (2006). Bonner observes the fixation with size as an indeterminate sliding and transformative force, but goes on to argue the necessary importance of size to biology. For Bonner, size places all kinds of limits on existence. He notes the disjunction when he observes that "organisms are material objects while size is a bloodless geometric construct."[32] While architecture is bloodless and often spoken of as a geometric construct, it remains material and, thus, bound to its own series of rules as to how size is accommodated and what its impact is.

The Centre of the Cell is the largest of the Blizard Building's internal pods, with a floor area of 2,100 square feet (195 square meters) on two levels. It has its own internal stair. The interior was designed by Land Design Studio, which specializes in expo pavilions, museum exhibitions, and visitor centers. Their promotional byline is "we tell stories in cultural and commercial spaces through the integration of interior design, architecture, scenography and communication media."[33] The story they tell here is of science—the science of cells, healthy and pathological. The Centre of the Cell had its own opening in 2009. Helen Skelton, a presenter on the world's longest running children's television program "Blue Peter" (CBBC) was given the privilege of cutting the ribbon, a choice suggestive of its intended audience. (By comparison, Her Majesty, the Queen, opened the Sainsbury Laboratory in Cambridge in April 2011 and the Francis Crick Institute building in London in November 2016.) The interior of The Centre of the Cell is overwhelming. One imagines even the most attention-seeking child would be quieter than this space. Its interior glows whatever color is decided upon and contains screens and projections of all types. It alone cost USD 3 million (GBP 2.5 million) and is larger in cost and floor area than the smallest laboratory in this book, Harlem Biospace (discussed in chapter 12). Land Design Studio proposes that the Centre of the Cell is engaged with "making the work of research scientists visible and more relevant to the lives of the young people."[34] Oddly enough, the scientists operating below are not visible from within the cell. The gap between stated objective and actuality here suggests a certain rhetorical quality to the biological analogies of the Blizard Building. It is not that the sight of scientists at work is supplemented or explained by the stories told here through education programs and biological analogies; it is substituted by them.

In *Beyond Living: Rhetorical Transformations of the Life Sciences* (1997), Richard Doyle updated Bonner's "Analogies in Biology," writing:

> The gap and border between DNA and proteins, numbers and words, codes and organisms is both the site of imprecision and the site of metaphorical intervention. The problem of "translating" life is one possible way of deciding on and effacing the border between textuality and vitality, a translation that appears within an episteme in which "Life becomes one object of knowledge among others," an object in and of language.... This imprecision of life seems to provoke a rhetorical crisis; each trope we deploy—code-script, translation, program—seems to provoke different conceptual blind spots, oversights that then render any account of living systems inadequate, imprecise.[35]

We can rephrase Doyle and note the gap and border between science, scientists, and consumers of science is also the site of imprecision and metaphorical intervention. When we step out from the Centre of the Cell, we see below us scientists in their open lab (figure 6.7). Not many—two or three. They prefer to occupy the blind-spots under the mews. This open lab is organized into rows of double-sided benches, spaced precisely, evenly, over the sky-blue floor. Each bench has an identical overhead lighting system covered by a bench-long rounded azure-blue lampshade. The long lampshades seem to serve two purposes. They bathe the benches in light and protect them from dust that might fall. From above, they serve to focus our attention on the scientists—they operate like theater lighting: the scientists are washed in light, and the lampshades shield us from possible glare. The same lampshades also obscure much of the contents of the benches.

It is rare to see scientists like this, from above, as one might watch a zebrafish wiggle in a petri dish. The fourth wall of this theater is gently marked. There's no proscenium arch, but rather a slim black band running horizontally around the periphery walls, forming an invisible floor that separates this laboratory and the spaces above. This line marks the complex ventilation system that secures the biological containment of the wet laboratory. Air comes in one side and is sucked out the other, guaranteeing any stray pathogens are removed before they leak into the void. This line also marks the separation between the spaces populated by the public, the pods, and the graphic panels, and the spaces occupied by scientists, benches, and actual cells. And it also marks a division between science as infotainment and science as profession. For students to move from one space to the other, they must be subjected to a series of institutional modulations, from A levels and university degrees to internships and postdoctoral fellowships.

The analogical molds that are occupied at the Blizard Building are carefully articulated: "Enclosures are *molds*, distinct castings," as Deleuze wrote. The areas that belong to scientists, children, and other visitors are precisely articulated containers, carefully connected and defined. Bonner suggests that in place of analogy we might note that "the environment both without and within the organism along with the inter-relation of parts of an organism and organisms one with another should all ultimately be described in terms of a complex scheme of interlocking activities or functions."[36] Further, what we will end up with is "a comprehensive catalogue of how living substances can be marshalled to perform certain jobs."[37] Bonner is not referring to the Blizard Building and its educational agenda, of course. He is describing the complexity that biological analogy often distracts us from.

Each of the elements that constitute the Blizard Building is carefully articulated. All the elements are "marshaled to perform certain jobs." Like a primary-school textbook, the Blizard Building involves a careful framing of the joyous and the educational, of access and control. "But controls are a *modulation*," says Deleuze, "like a self-deforming cast that will continuously change from one moment to another, or like a sieve whose mesh will transmute from point to point." The near seamlessness by which we move from a Centre of a Cell, to a bridge, to an amphitheater, to the mews, and back out onto a street in Whitechapel is dazzling. Analogy is the theoretical lubricant. Its corollary is "lifelong education." Here, at the Blizard Building we go seamlessly from the cell, the microscope, and benchtop to the mushroom, the street, and the classroom.

7 Excess

Biosciences research takes place on and through the bodies of animals, be they the transparent nematodes and zebrafish or mammals with which we share much DNA. Animals are a critical part of experimental biosciences research. The spaces in which animals are bred and held in pathogen-free colonies are called "vivaria." Typically, vivaria are hidden from public view, at the peripheries of buildings, the fringes of campuses, or in remote locations. There are animals concealed in the basement at NavarraBioMed (2011) in Pamplona, Spain, designed by the Spanish firm Vaillo and Irigaray Architects (figure 7.1). The building has a total floor area of 85,000 square feet (7,854 square meters), and a large portion of one of its four floors is dedicated to housing animals, predominantly rodents. Despite this placement of real animals, the animal is curiously memorialized in this building, albeit in such a way that it would not recognize itself. The animal is key to architectural expression here. The architects propose the building is analogous to the camel, the polar bear, and the leaf. Its decorative beauty and animal analogies are unusual in the architecture of biomedicine, while its use of animals in experimentation is commonplace. The engagements made with "real animals" at NavarraBioMed are not exceptional; and we have not selected it as a case study building because it uses more or less animals in experimentation than any other similar organization. Its selection is, instead, because its architecture expresses the complexity and excesses of the animal in laboratory architecture.

Architecture has a long history of engaging in obscure intimacies with the animal. But perhaps this engagement has been far too obscure. Since Vitruvius, reterritorializations of architectural form as organism, as body, or as receptor of bodily geometry have flourished.[1] The human form has tended to dominate, but the presence of the nonhuman animal has grown. In *Changing Ideals in Modern Architecture, 1750–1950* (1965) Peter Collins traces the era of "The Biological Analogy" to around 1750 and Georges-Louis Leclerc, Comte de Buffon's monumental forty-four volume publication *Histoire*

Figure 7.1
Exterior view of the main entry to NavarraBioMed (2011) in Pamplona, Spain. Architects: Vaillo
and Irigaray Architects. Photograph by Jose Macutillas.

naturelle (1749–1804).[2] It is the era when the French architect Jean-Jacques Lequeu
etched capricious proposals such as "The Cow Stable faces south on the cool meadow"
(1790), a folly resembling a cow, complete with anatomically correct horns, eyes,
ears, legs, and hair, and adorned by a bell, urn, and blanket. Lequeu proposed, too, an
inhabitable edifice taking the form of a giant elephant (1758) and an equally large and
flamboyantly attired sheep (1789). A century later, René Binet's *Esquisses décoratives*
(1900) contains a sequence of architectural designs based on the biological illustrations
drawn by the zoologist and morphologist Ernst Haeckel.[3] Here, too, there is a shift in
scale and function, but it is even more extreme. Envisioning tiny shell-like skeletons
as monumental architectural structures, Binet scaled Haeckel's microscopic biomineral
creatures, or *radiolaria*, into decorative façades, trellises, and follies. Binet's architectural
translation of Haeckel's lithographs culminated in the design for a Monumental Gate
at the eastern entrance of the Paris Exposition Universelle in 1900 (figure 7.2). Binet's
"Porte Monumentale" is based on one radiolaria illustration alone—the *Cyrtoidea Ptero-
canium trilobum*. Binet wrote to Haeckel explaining that "everything about it, from
the general composition to the smallest details, has been inspired by your studies."[4]
However, there is evidence to suggest Binet was engaged in a process of abstraction and

Figure 7.2
René Binet, Monumental Gate at the eastern entrance of the Paris Exposition Universelle (1900).
Photograph courtesy of the National Gallery of Art, United States.

geometric idealization influenced by the arabesque aesthetics of art nouveau and the conviction that nature exhibits geometric order.

The architectural historian Geoffrey Scott was perhaps the first to notice the confluence of the images and languages of biology and architectural theorization.[5] In a chapter tellingly titled "The Biological Fallacy," Scott notes the appropriation of the language and understandings of evolutionary biology in his work on *The Architecture of Humanism* (1914).[6] Philip Steadman hones in on analogical engagements with the animal in *The Evolution of Designs: Biological Analogy in Architecture and the Applied Arts* (1979), which we have already discussed in chapter 6. Analogy presents itself, however, as a particularly restricted economy. In *Architecture, Animal, Human: The Asymmetrical*

Condition (2006) Catherine Ingraham asks us to imagine a future when animals "have the power to completely alter our way of thinking about ourselves, both the form of ourselves and the forms we make, live in and respond to."[7] Ingraham laments that "imagining a paradigm shift in architecture in which animal life would play a part is an extremely difficult task."[8] For Ingraham, too, the engagement with the animal occurs as a type of thinking and imagination related to architectural expression, even as she attempts to move beyond anthropocentric accounts. However, sometimes engagements between architecture and the animal are far more intense than thinking, and relate more to consumption.

The debt both science and architecture owe animals goes largely unacknowledged, even in the case of NavarraBioMed, where humps and fur animate the building's silhouette and surfaces. It is a debt at once material and semiotic. A complex *economy* is to be found here. It is not an economy based on balanced spreadsheets, although the language regulating the use of animals in scientific experimentation is drawn from accountancy: "The likely harm to the animal should be balanced against the expected benefits of the project."[9] This economy operates in futurity. That is, investments—of money and effort, as well as animal and human lives—in the biosciences are always framed as investments in the future. Or, as stated in the 2010 European Parliament Directive (Directive 2010/63/EU) on the protection of animals for scientific purposes, "Their use in procedures should be restricted to areas which may ultimately benefit human or animal health, or the environment."[10] These investments (expenditures) are framed as being absolutely necessary, but the rewards (productions) are *ultimately* deferred. Indeed, the rewards are deferred and *displaced*, as they accrue to those who made no sacrifice. An example would be the sacrifices of nonhuman primates whose use in biomedical research in Europe is permitted only for "the avoidance, prevention, diagnosis or treatment of debilitating or potentially life-threatening clinical conditions in human beings."[11]

Pamplona is, of course, famous for another "frenzied expenditure" and "act of sacrifice" centered on the animal, the Festival of San Fermin, the running of the bulls. On a hot summer day during the Festival San Fermin, NavaraBioMed shimmers. The building is part of the Navarra hospital campus, not far from the old city and its famous Plaza de Toros. Like most older hospital campuses, it has an eclectic mix of architectural styles from many periods, as well as the usual accretions and piecemeal interventions that adapt older buildings to constantly changing demands. In this context, NavarraBioMed is singularly coherent—a free-standing three-story building (plus basement) with a perfectly rectangular plan. Against other new laboratory buildings, it is a particularly elegant example, glamorous even. The orthogonal shifts of its stepped volumes are precisely and asymmetrically balanced. It is wrapped in a shimmering folded façade of perforated metal, which its architect tells us is "a shell that covers the characteristic forms. The outer skin 'traces' the internal structures."[12] It is a dazzling jewel. It is one of the few laboratories where the external image of the building is of such iconic character that it is conspicuous on almost all the marketing and communications coming out of NavarraBioMed. Even financial statements are emblazoned with its image. It is a laboratory whose beauty is excessive.

To aid the task of exploring the economy of the laboratory, and of NavarraBioMed in particular, we will introduce the first volume of Georges Bataille's three-part series, *The Accursed Share* (1967) and the distinction he makes between a *restricted economy* and a *general economy*. Restricted economies are concerned with notions of scarcity and balance and relate to accountancy and scientific analysis. Bataille's interests lie, however, with the necessary excesses and surplus energies that put such economies into motion. According to Bataille, the "accursed share" is excess (*dépense*) spent lavishly and knowingly without obvious benefit. Examples include the expenditures of frenzied energy of predatory animals in pursuit, nonreproductive acts of sex, extravagant spectacles, and destructive acts of sacrifice. Bataille argues for the necessity for a society to sacrifice, or "burn off" excesses as "excess energy, translated into the effervescence of life."[13] Through Bataille we can explore the consumption of animal lives, both analogical and real, which lubricate the operations of the contemporary laboratory.

Economies of Excess

Bataille commences *The Accursed Share* with the line "Exuberance is beauty." The quotation comes from the poem "Proverbs of Hell," published in *The Marriage of Heaven and Hell* (1790) by the eighteenth-century English artist and mystic William Blake. Against a theological background, the poem argues the necessity of investments that are made in respect to reason and passion, courage and cruelty, and excess and sacrifice. Blake draws upon a vast range of images. He invokes animals—"the rat, the mouse, the fox, the rabbit: watch the roots; the lion, the tyger, the horse, the elephant, watch the fruits"—as well as architecture— "Prisons are built with stones of Law, Brothels with bricks of Religion." Blake also invokes the dead: "Drive your cart and your plough over the bones of the dead. The road of excess leads to the palace of wisdom".

Under Blake's banner, "Exuberance is beauty," Bataille reignites excess and sacrifice using these same points of fixation: animals, architecture, and the dead. Bataille rejects traditional and habitual notions of economy as "limited." He bemoans that "economic activity, considered as a whole, is conceived in terms of particular operations with limited ends."[14] Instead, his ambition is to imagine a "'general economy' in which the 'expenditure' (the 'consumption') of wealth, rather than production, was the primary object."[15] To do so, Bataille extends the question of economy to "the play of *living matter in general*."[16] Abandoning the usual strictures of economic analysis, Bataille's account roams across literature, poetry, architecture, history, biology, anthropology, sociology, sexuality, and politics. In place of an exploration of balance, stability, and trade, this wide embrace leads to an exposition on excess. The manners by which excess is dealt, and the various relegations of positive or negative excess, is a focus of *The Accursed Share*. Excess is "accursed" in the sense it must be dealt with, lest it deals with us: "For if we do not have the force to destroy the surplus energy ourselves, it cannot be used, and, like an unbroken animal that cannot be trained, it is this energy that destroys us; it is we who pay the price of the inevitable explosion."[17] This issue is resolved in the "burning" of excess energy. What is habitually framed as waste and loss is considered by Bataille as a necessary potlatch for the ongoing existence of a society or a species. Whereas ancient societies held festivals or erected purposeless monuments, "We use excess to multiply 'services' that make life smoother, and we are led to reabsorb part of it by increasing leisure time."[18] Aesthetics and consumption, glamor, games, and entertainment are implicated in this logic of excess, which operates against stability and utility. Architecture, too, is implicated. However, before turning to the specificities of biomedical laboratory architecture, it is worth considering the role of the animal in Bataille's narrative.

Bataille introduces much of his thesis by referring to domestic animals—the calf, the cow, the bull, the ox, dog teams—and to "unbroken" and wild animals—the human, and the tiger.[19] French anthropologist and ethnologist Claude Lévi-Strauss had suggested a decade earlier that the role of the animal in totemic rituals was important "not because they are good to eat, but because they are good to think."[20] Such an understanding might be extended more broadly to the human-animal correlation. Blake's and Bataille's specific engagement with the tiger is an example of operating *in thought* through an animal. A few years after the publication of *The Marriage of Heaven and Hell* Blake wrote of the beautiful exuberance of "The Tyger" in *Songs of Experience* (1794): "Tyger Tyger, burning bright, / In the forests of the night; / What immortal hand or eye, / Could frame thy fearful symmetry?"

The image of the tiger drawn by Blake is perplexing (figure 7.3). Though obviously muscular, this tiger is not particularly terrifying. Its teeth and claws are withdrawn, and it looks somewhat dazed. Despite the divergence in textual accounts, Blake's tiger does bear similarities to the etching of a tiger in Buffon's *Histoire naturelle*.[21] Both are viewed from the side as they walk in a clearing. The Blake scholar John E. Grant notes that

Figure 7.3
William Blake, "The Tyger," plate from *Songs of Experience* (1794). Courtesy of the British Museum, public domain.

a tree runs up the right-hand side of the page and extends its branches horizontally into the poem. In some versions it is striped to suggest "it is the vegetal equivalent of the Tyger."[22] Blake implies a relation between animal, vegetation, page, and text that extends the theological questions as to the force of creation well beyond the animal toward "the play of *living matter in general*," to use Bataille's phrase.

In a section of *The Accursed Share* titled "The Three Luxuries of Nature: Eating, Death and Sexual Reproduction," Bataille turns explicitly to Blake's tiger and from the *thinking* of the animal toward the more visceral characteristic of *eating*. Bataille notes, "The eating of one species by another is the simplest form of luxury"[23] and in a food chain can be defined in terms of excess. "William Blake asked the tiger: 'In what distant deeps or skies burned the fire of thine eyes?' What struck him in this way was the cruel pressure, at the limits of possibility, and the tiger's immense power of consumption of life. In the general effervescence of life the tiger is a point of extreme incandescence."[24] For Bataille, the tiger comes to represent a gratuitous excess of speed and consumption. The tiger consumes enormous resources and burns incredible energies in its quest for satiety. The tiger is luxurious, wild, and intense. The tiger burns bright, and its incandescence is brilliant and explosive. It is a type of excess to which life, and lives, are subject even in the face of measures that might seek to control and regulate. For Bataille, real excess commences when the individual, or group, has stopped growing—or reached its limit.[25] He goes on to note that Blake's account of the tiger operates at the limits of possibility. Bataille describes this moment in a life of an individual, society, or species as "always bordering on explosion"[26] and focuses on such moments saying:

> Sexual reproduction is the occasion of a sudden and frantic squandering of energy resources, carried in a moment to the limit of possibility (in time what the tiger is in space). This squandering goes far beyond what would be sufficient for the growth of the species. It appears to be the most that an individual has the strength to accomplish in a given moment. It leads to the whole-sale destruction of property—in spirit, the destruction of bodies as well—and ultimately connects up with the senseless luxury and excess of death.[27]

Thus, Bataille entwines matters of the flesh with conceptual concerns; titillation with rumination; the visceral with the incorporeal. With the aid of Blake's tiger, Bataille extends Levi-Strauss's position on the totemic engagement with animals. Bataille suggests that the animal is as necessary for matters of content as it may be for matters of expression. The tiger allows Bataille to weave a narrative of consumption, excess, sex, and death. This logic of excess falls into the realms of Charles Darwin's 1871 account of sexual selection as a form of nonfunctional virility. A logic of excess operates in terms best described as "in spite of." In spite of the restricted economy and its measures, the tiger exists. The tiger exists in spite of its immense consumptions and in spite of its flamboyant coloration. It is this luxurious flouting of the measures of balance and stability that is itself brilliant in the pink of the flamingo, the fan of the peacock, the gold crucifix on the chest of a drink's waiter, and the blazing stripe of the tiger. The force of this nonfunctional logic, when considered to be without boundary or restraint, is entrancing. Bataille extends the logic from the animal to the human, from sex to death, in time and in space, and onto the inorganic and the play of living matter in general and the relations of all living matters, celebrated or sacrificed. One thinks of Blake's illuminated transcript and Levi-Strauss's account of totemic societies when Bataille notes his personal nostalgia for "a time when the obscure intimacy of the animal was scarcely distinguished from the immense flux of the world".[28]

The Restricted Economy of NavarraBioMed

As we have seen, the biomedical research laboratory has become one of the most sumptuous investments of the twenty-first century. The level of investment is comparative to that made in submarine technology during World War I, aircraft technologies during World War II, or the space race made during the Cold War. There is a consistency to the rhetoric behind such investments: we fail to invest at our own peril; we are told our future is at stake. This rhetoric pervades the biosciences and is most acute in biomedicine where serious levels of investment are necessary to cure cancer, prevent epidemics, slow aging, and so on. The expensive buildings that house this research have proliferated across the period of the twenty-first-century global financial crisis. They emerge as cathedrals did, in spite of the restricted economy of the Middle Ages. Such a level of investment either represents a massive squandering of resource, a sincere investment in the future—or both. In chapter 10, we interrogate the complex role of philanthropic investment in science as a form of potlatch or redress to excesses of wealth. In chapter 11 our focus moves to the speculative role of angel investors, but for NavarraBioMed, the "general economy" is a complex mix of bodies and buildings, and state and private investment that deserves untangling.

Let us first examine its restricted economy. The restricted economy of NavarraBioMed laboratory is easy to account for. In 2007, the Health Department of Spain approved a "Strategic Health Research Plan for Navarre,"[29] which centered upon a perceived need to increase critical mass in biomedical research. It's a plan that to date is unprecedented in the Navarre region, and Spain more widely. Countries internationally were (and are) making serious investments in the biomedical arena and particularly in modes of translational medicine, which encompasses medical research, clinical care, and pharmaceutical production. The "Strategic Health Research Plan" also deals with some of the communication and representational issues related to biomedicine and the need for public support for large-scale state investment. The Navarre Plan specifically seeks to "improve the systems of communication and drive the transfer of knowledge and technology to the industrial production sector."[30] The industrial production sector in this context is primarily the pharmaceutical industry. However, there is little detail as to what would be disseminated to the public.

NavarraBioMed is an outcome of the "Strategic Health Research Plan for Navarre." NavarraBioMed was previously called the Biomedical Research Centre of the Health Service of the Government of Navarre (Centro de Investigación Biomédica del Servicio Navarro de Salud del Gobierno de Navarra), also known as Centro Investigación Biomédica (CIB). The development of NavarraBioMed is framed as a strategic tool to advance the development of the "Plan Moderna," or "Modela de Desarollo económico para Navarra."[31] NavarraBioMed is implicated (if only rhetorically) in the region's economic development and with the general task of spreading enthusiasm for, and investment in, biomedicine. Its architecture is very much part of this project. The European Investment Bank, which provided a loan of USD 317 million (EUR 300 million) to finance projects in the Navarre region did so in the belief that "in its new building and with a strategic plan in motion, the Research Centre will be a benchmark in biomedical research in Navarre."[32]

The architects, Antonio Vaíllo i Daniel and Juan L. Irigaray Huarte, were commissioned by the Navarre Health Service (Servicio Navarro de Salud—Osasunbidea) to design the NavarraBioMed building following a restricted competition.[33] The architects had designed a cardiology pavilion for the same client on the Navarre Hospital Campus. The NavarraBioMed building was completed in 2011 at a purported cost of USD 19 million (EUR 18 million).[34] Though relatively inexpensive for a biomedical research laboratory, it is likely the building was economically straining in the context of Spain's experiences of the global financial crisis.[35] Its opening and operation were staggered due to the economic situation. Operating alongside CIB since October 2012 is the Miguel Servet Foundation, and the two organizations are collectively are known as NavarraBioMed. The aim of the Miguel Servet Foundation is to promote healthcare research; coordinate programs and facilities; and maintain the research teams and healthcare professionals of the Health Service of Navarre. The foundation is named after the sixteenth-century Spanish theologian, medic, and scientist, who was immolated for heresy.

The key areas of investigation of NavarraBioMed are in epidemiological surveillance; public health and health services; mental health; neuroscience; oncology; and the pathology of large systems. In addition to funds from the European Investment

Bank, NavarraBioMed operates with funding from the European Union, Fundación Caja Navarra, the University of Navarra, the Public University of Navarra, the Obra Social "la Caixa" Foundation, ADItech (Advanced Innovation and Technology Corporation), the Instituto de Investigación Sanitaria de Navarra, the Gobierno de Navarra (Department of Health, Government of Navarra), and a supportive tax regime. The objective of the Navarre Plan to "drive the transfer of knowledge and technology to the industrial production sector" seems to have been fulfilled.[36] Strategic links between NavarraBioMed and Novartis Pharmaceuticals have already been established. Alongside Novartis, the University of Cambridge and the Spanish Ministry of Economy and Competitiveness are partners. The Irache funeral home and the Morgue San Alberto are noted for their contribution to the NavarraBioMed biobank of neurological tissue (brains) and tumors.[37] Publicly noting such contributions from organizations so closely connected to human death is unusual for a biomedical research laboratory.

The General Economy of NavarraBioMed at the Limits of Possibility

The general economy of NavarraBioMed is harder to explain than its restricted economy. Its decorative beauty and its engagement with ideas that derive from animal analogies are at once alluring and perplexing. We have noted that the use of "real animals" at NavarraBioMed is not exceptional. Indeed, compared with the 20,000-square-foot vivarium within the 2-acre basement serving Weill Hall at Cornell University, NavarraBioMed's hidden floor is modest.[38] It does, however, express some of the complexity and excess of the general economy. It is unapologetically exuberant, and the use of exotic animals for the purposes of architectural biomimicry may parody the extravagant expenditure of the less exotic rodents that live and die within.

While its rectilinear layout is highly functional, its surfaces and silhouettes are expressive, its finishes luxurious, and its textures ornamental (figure 7.4a and 7.4b). Inside, circulation spaces are printed with images from microscopy, digitally enlarged and graphically transformed into shimmering metallic surfaces and acidic colors. A meeting hall is lined with floor-to-ceiling blood-red crushed velvet drapes on all sides, a theatrical and decadent gesture at odds with the pure white drapes seen in Fabrikstraße 22 in chapter 5. Externally, the volumes and disposition of the parts fall beneath a skin that does not recognize any hierarchies between parts. The exterior screen is not a thinly applied pattern, but in the tradition of relief sculpture, has a volumetric quality that performs an agitation and thickening of the surface. How might we understand the excesses of architectural ornamentation and analogy in relation to expenditure and sacrifice?

Figures 7.4a and 7.4b
Decorative surfaces inside NavarraBioMed (2011), Pamplona, Spain. Architect: Vaillo and Irigaray Architects. Photographs by Jose Macutillas.

The very word "pharmacy" is etymologically connected to sacrifice. The *pharmakos* in Ancient Greece was the ritualistic sacrifice or exile of a human scapegoat. A slave, cripple, or criminal was chosen at times of disaster or crisis in the belief their sacrifice would bring about purification. On the first day of the festival of Apollo at Athens, two men, known as the *pharmakoi*, were sacrificed as a form of reparation. They were, according to Walter Burkert, first fed well—a precursor to the customary North American ritual of giving prisoners on Death Row a last meal of their choosing preceding their execution.[39] The shift from human to animal sacrifice in pharmacology mirrors ideas around sentience and cruelty. Today, we sacrifice animals, burn through Euros, and burn out scientists in order to secure the future. Bataille notes, "One no longer imagines that cruelty can seem unavoidable. But this world of ease has its limits. Beyond it, situations arise in which, wrongly or rightly, acts of cruelty, harming individuals, seem negligible in view of the misfortunes they are meant to avoid."[40]

The belief that animals lack consciousness and do not experience pain and suffering as human beings do was a keystone to the philosophy of reason argued by René Descartes in the seventeenth century. Accordingly, this era saw a wide range of very invasive experiments, of which Robert Hooke's on artificial respiration and William Harvey's on the circulation of blood are the best known. There was not yet a dialogue around animal rights or protections. In this context, and in the absence of anesthesia, the ideal laboratory animal was the sheep, given its placidity and availability. This changed with anesthesia, wherein all kinds of animals could be made placid. "Beginning in the 1870s," Jed Mayer writes, "animal experimentation came to be more widely practised in British physiological laboratories, bringing into conflict scientific and humanitarian interests in the animal, as anti-vivisectionists and advocates of animal research debated their right to speak for the nonhuman animal."[41] This period saw the emergence of public zoos and animal protections societies alongside the popularization of Darwin's theory of natural selection, which charted the relations between all animals (including the human). It also saw the first legislation directed toward the protection of animals used in science: the 1876 Cruelty to Animals Act in the United Kingdom. In the United States, there were no restrictions to experimentation on animals in laboratories until the passage of the Laboratory Animal Welfare Act in 1966. Disturbingly, the act covered only facilities that used domestic pets: dogs and cats. The choice of animal, until the 1960s, related to its approximation to human physiology, hence the popularity of primates, and the capacity for standardization of the animal-subject. Primates continue to be used for research into drugs and vaccines, although chimpanzees have not been used for experimental purposes since 2000. Mice, rats, fish, and birds accounted for 93 percent of research animals in Spain in 2014, roughly the same proportion as in other EU countries.[42] Dogs, cats, and primates—the flashpoint species for activists—account for less than 0.2 percent of all research procedures. Today, the principle of minimizing distress and pain that guides the regulation of animal use in European laboratories aligns with the notion that animal species have different *capacities* for experiencing pain. Even so, there are clauses permitting "the use of a procedure involving severe pain, suffering or distress that is likely to be long-lasting and cannot be ameliorated" based on "exceptional and scientifically justifiable reasons."[43] The nonhuman animal continues to be viewed not as an individual with social relationships and rights equal to the human, but rather as a kind of anonymous sacrificial knowledge generator.

In spite of the shifting volumes and the shimmering façade of NavarraBioMed, we notice that in the shadows the ground plane itself folds down to allow access to a basement level (figure 7.5, figure 7.6). The laboratories we have visited often contain such a level. Indeed, the laboratory was born here—below ground—in the basements of the alchemists. In an era when public investment implies a reciprocal public access to laboratories—albeit visual access—we should note that not every aspect of biomedicine has risen from the basement or finds itself transparently expressed. Access is tightly restricted to these basements, and images are not easy to take or to find. In all our travels to and through biomedical research laboratories internationally, we have been consistently and politely denied access to the vivaria. Laboratory protocols are invoked as the key reason, often with an accompanied suggestion that not entering is in the best interest of the animals occupying these basements. Plans from the architects, too, are vague

Figure 7.5
Exterior of NavarraBioMed (2011), Pamplona, Spain, from the service lane on the east side, from where you can just make out the basement undercroft. Architect: Vaillo and Irigaray Architects. Photograph by Jose Macutillas.

Figure 7.6
Close-up view of the basement of NavarraBioMed (2011), Pamplona, Spain. Architect: Vaillo and Irigaray Architects. Photograph by authors.

on these areas. (Access to the animal house at one of the translational research buildings we visited was through a locked door whose signage declared it was a universally accessible toilet. Should an unauthorized visitor penetrate the security lines to this private floor and those of its locked toilet door, he or she would find, indeed, an accessible toilet. From there, another locked door leads to the animal house.) NavarraBioMed's basement level contains a large vivarium (some 6,800 square feet, or 632 square meters), which is referred to as a "living area" for the animals, including specifically pathogen-free (SPF) animals, along with some technical/servicing and general storage. There is also an experimental surgery located in the basement. There are no windows down there. It's not that light would cause a problem—but sight is sometimes a dangerous thing.

The invisibility of this important part of the research scene is itself of interest. Discussions of connectivity and social space and the expressive potential of scientific operations do not extend to the animal populations, the spaces they inhabit, and the experimentation to which they are subject. However, the ways animals respond to the laboratory as an environment is an area that has been much studied since the 1960s.[44] According to Paul Wuldau of Tufts University, animal research laboratories are made invisible to protect proprietary knowledge and for defense against animal rights activists.[45] This is a distinctly modern concern. Opponents to the use of animals in scientific research have a range of concerns: safety and rights of the animals being tested and safety for the environment if animals with disease or genetic modification are released. (Some of these animals are genetically engineered for the purposes of better modeling human diseases and to test for chemical toxicity.) We noted in chapter 3 that biological dangers occupy either side of the walls that demarcate the laboratory. Arguments in favor of animal testing, as we have seen, include the potential medicinal, nutritional, and conservation benefits, the need to maintain the research of each nation in the global marketplace, and the rights of researchers. But for research centers that use animals it is far easier not to have to make these arguments at all. The discourse of the laboratory—even that discourse related to the animal—emphasizes healing and cutting-edge research, rather than animal experimentation. In *The Sacrifice: How Scientific Experiments Transform Animals and People* (2007), Lynda Birke, Arnold Arluke, and Mike Michael note that the "passive voice [of scientists] is coupled with euphemisms and the frequent omission of details of how animals live in laboratories, which combine to obscure what happens to animals. There are various euphemisms, but the most obvious is the word 'sacrifice' rather than 'kill' in scientific reports, drawing parallels with ritual sacrifice. This gives killing a symbolic importance, as though the animal were sacrificed for some greater good."[46]

NavarraBioMed is built, as many biomedical laboratories are, on the lives of animals. Bataille documents the sacrificial character of preindustrial societies—Aztec, Mexican, pre-Islamic Arab, Islamic, and Tibetan—to trace the necessary economy of consumption and sacrifice. For the Aztecs, the "science of architecture enabled them to construct pyramids on top of which they immolated human beings."[47] Bataille points to the care and respect paid to the sacrificial victims, which lubricated the general economy of past societies. The animals of the NavarreBioMed are likely well treated[48]—not quite as well as the *pharmakoi* of Ancient Greece or the Mexican sacrificial victims of Easter, but well nevertheless.

Denis Hollier notes Bataille's anxious relation with architecture as a system of expenditure that operates on establishing a "difference between high and low, between dirty and clean ... slaughterhouses and museums."[49] This difference occurs spatially in NavarraBioMed, and many of the laboratories we observed, in the division between animal houses and wet laboratories. The animals are born in the "living area" and many die there, or in the laboratories. Ingraham turns to the point in her book *Architecture, Animal, Human*: "The slaughterhouse/museum 'system' ... is also a system that brings animals, human beings, buildings, and the coalescence of buildings and infrastructures into cities, into entangled relations with one another. Animals are, in effect, spent in the slaughterhouse in order to fuel the economic and biological transfer of energy from animal to human."[50]

Birke, Arluke, and Michael suggest that laboratory animals "ambiguously signify both the selfless biomedical enterprise, which has putatively yielded so many benefits, and the perceived hubris and arrogance of scientific institutions and scientists."[51] Architects are not immune from this same hubris. The real animals "spent" in the basement are perhaps of even less concern to the architects of biomedical research facilities, who are concerned instead with what Ingraham refers to as the "symbolic and conceptual transfer" of the animal.[52] The architects of NavarraBioMed would seem to concur with Lévi-Strauss in engaging animals "not because they are good to eat, but because they are good to think." The animals with which the architects think are bigger and more exotic than the mice, nematodes, fruit flies and zebrafish of experimental science. The architects of NavarroBioMed write anatomically of this monument to science:

> The look is inherent in this intrinsic functionality. It shows up through one enclosure that covers all of its formal determinations. The outer skin traces its internal structures. The camel, the polar bear and the leaf: the project aims to link with the content of the program: Bio-Medical Research, through the implementation of the biomimicry (adaptation of biological systems and human devices) in the process of architectonic generation.[53]

The architects undertake to conceive "the world as a system of inter-related systems."[54] Biomimicry is engaged as the device through which to link the architecture of the laboratory to its biomedical content. However, it is perplexing that such mimicry in the north of Spain has invoked camels and polar bears (figure 7.7). Perplexing, too, that the place in this biomedical facility for the real animal is in the basement. In lieu of exploring the basement, the architects engage *thought animals*—fabled, exotic, and wild animals, animals that do not bleed, procreate, or die. One critic of NavarraBioMed writes, "They take these three bio-reference types to achieve similar adaptive systems: The camel as a paradigm for functional section; The polar bear skin as an example of multifunctionality; The leaf as integration between structural resolution and flexibility."[55] Function and structure are consistently identified in such accounts as the reason for the biomimicry. The architectural critic is concerned with the restricted economy of production rather than the general economy of consumption.

The articulation of the profile of the NavarraBioMed building breaks the volume into seemingly four parts, with each section marked by the increased volume of roof. The mechanical services are accommodated in rectangular blocks located on the uppermost level (roof) of the building in structures that are likened to the humps of a camel: "The architects themselves compare the longitudinal section of the building with a camel whose body shape is embossed with the characteristic bumps on the requirements of the function."[56] It was a strategy Vaillo and Irigaray employed in their earlier *b3* house project in Pamplona during 2009–2010. In this project the same formal maneuver is not articulated in biomimetic terms.

Figure 7.7
Elevation of NavarraBioMed (2011), Pamplona, Spain. Architect: Vaillo and Irigaray Architects. Photograph by Jose Macutillas.

The analogy of the polar bear is even more obscure. "The image is inherent to its intrinsic functionality," says architect Juan Irigaray Huarte. "The 'skin' of the CIB is an imitation of that of a polar bear; in extreme cold weather conditions, the skin maintains stable internal temperatures. The perforated skin of CIB building also echoes origami with its folding and repetition."[57] While one may imagine a flamingo performing origami, a polar bear is unlikely to do so. NavarraBioMed has floor-to-ceiling glazing behind the folded silvery-gold anodized aluminum sheets ($177 \times 31 \times 1/8$ inch, or $4,500 \times 800 \times 3$ millimeters). The origami surface serves as solar protection, but more importantly reduces views into the building and gives the building its appearance as a singular object. The discussion of the screen of NavarraBioMed in the text of architecture journals tends to fixate on the "intrinsic functionality" of biological structures, as if animals were highly refined machines, with every part, every color, stripe, and plume functioning efficiently and economically—without waste, without excess, without pleasure. But much of biology, like much of architecture, is a type of nonfunctional anatomy. The flaming stripe of a tiger is expressive and excessive and necessarily so. The screen of NavarraBioMed has nothing to do with a polar bear and neither is it concerned with functionality. The screen is an exuberance. It's not that the screen doesn't "screen." It does, but function is not the primary aim.

As Antoine Picon has argued, historically the degree, size, extent, and workmanship of ornamentation in architectural history relates to the perceived economic and social value of the building or its occupant—the richer the investment in functionally superfluous ornament, the more prestigious the building.[58] Because of its production costs, ornamentation expressed the economic power of the client and in doing so, signified their political power. The status of ornament shifts in modernism—from Alois Riegl's belief in ornament as the collective expression of the soul of a people to Adolf Loos's conviction it represented the worst of social decay and individual criminality. Its reappearance in architecture in the last thirty years decouples ornamentation from communication, but maintains its affective qualities. As Jonathan Massey bluntly puts it, ornamentation "makes things special."[59] There has been a revival of the idea, originally formulated by Gottfried Semper, that ornament expresses a vital force necessary for all living beings.[60] Ornament is not that which can be discarded without damage to an organism's functioning; rather, its excessiveness makes living possible. Human animals, too, partake and their "admirable monuments" can quite joyously have "no useful purpose." Exuberance is, after all, beauty.

8 Deep Time

The University of British Columbia's Pharmaceutical Sciences Building (PSB) in Vancouver was completed in 2012 at a cost of USD 98.4 million (CAD 133 million). The design by the Quebec-based firm Saucier and Perrotte Architectes, in association with local Vancouver firm Hughes Condon Marler Architects, is deceptively restrained in its external presentation to the street and carpark. It takes the form of a six-story-high rectangular glazed prism (400×165 feet; 120×50 meters). Above its slightly recessed ground floor, three of its elevations are smooth, reflective glass curtain walls. The east elevation, facing into the campus, is more articulated with a series of unevenly stacked, cantilevered room-sized glass boxes—an architectural motif that harks back to Moshe Safdie's Habitat housing (1967) for the World Exposition in Montreal and which was revived by architects MVRDV for their Wozoco Osdorp (1997) seniors' housing scheme in Amsterdam. The approach hints at prefabricated or modular construction, but as in most instances of its contemporary use, is not. This one photogenic elevation features strongly in the publicity for the project (figure 8.1). Viewed from the other three sides, including the side facing the main street bordering the campus, there is little to distinguish the PSB from the mass of generic, medium-rise buildings that make up office environments across the industrialized world. The PSB is much finer in its execution than most office buildings, but from nowhere outside is the shape and organization of internal spaces or the nature of their occupation revealed. Unlike the Blizard Building's façade, that of the PSB is not embellished with motifs drawn from pharmaceutical science as an indication of the activities taking place in its laboratories. It is, like Fabrikstraße 22, a building whose interior is more revealing of the ambitions of the architects and the organization than is its exterior.

The PSB's reticence in declaring its purpose stands in marked contrast with its campus neighbors completed the same year, the Earth Sciences Building (ESB) by Perkins and Will and the Beaty Biodiversity Research Center and Aquatic Ecosystems Research Laboratory by Patkau Architects. The ESB and Beaty Biodiversity Center both pursue a goal of "science on display," which is espoused by the University and linked to its

Figure 8.1
Exterior view of the Pharmaceutical Sciences Building (2012) at the University of British Columbia, Vancouver, Canada. Architects: Saucier and Perrotte. Photograph by Marc Cramer.

teaching mission. The ESB has a glass-fronted, double-height research laboratory, which, because it is sunken one floor below ground level, allows an elevated view of its activities from the vantage point of its exterior colonnade. One section of its façade is faced with twenty-one massive polished stone slabs of different origin installed in a decorative application called "book matching." The museum of the Beaty Biodiversity Center features a glass lantern-like, double-height entry space housing the suspended skeleton of a blue whale. This space, too, is sunken below ground level, so that the whale can be viewed at eye-level from the university's central pedestrian mall, and from below once inside. The skeleton is a visual lure and a pronouncement of the building's purpose—albeit it is also a suggestion of a species demise that the center might prevent.

Inside the Pharmaceutical Sciences Building

Against the orthogonal ordering and reserve of the PSB's exterior, its interior comes as a surprise. Much of the ground floor is dramatically cave-like. The foyer and atrium are defined by folded and faceted trapezoidal planes that lean into the space or wrap around stairs and corners. The ceiling and some of the wall planes are lined with Western red cedar boards, while others are of dark mirrored glass. The foyer is crisscrossed by narrow cracks of recessed fluorescent lights that exaggerate the discontinuity between and fragmentation of its defining planes. On the ground floor, the path from the east to the south entrance is oblique; there is no view through the building from one entrance to the other. The entrance to the main lecture hall is via a raked monochromatic tunnel. Every surface of the tunnel is a bright sunflower yellow, conveying the impression that the passage is cut through solid matter. Above the ground floor, the laboratories and offices are organized around two distinct top-lit atria, defined by the bands of Western red cedar lining the balustrades (figure 8.2). One atrium, between two zones of laboratories, provides light and access. The other divides the middle section of laboratories from a zone of offices and meeting rooms (figure 8.6). Both are irregular in section and bridged by ramps painted in contrasting black. The southern circulation space, between the laboratories and offices, is a long four-story-high slit just wide enough for two scientists to comfortably pass each other. It, too, is lit from above. Together, these elements comprise an interior of dramatic chiaroscuro and alternating compression and release.

The PSB houses the faculty's research laboratories for drug discovery and health care innovation: the Centre for Drug Research and Development. Auxiliary spaces include classrooms and lecture halls; an exhibition space conveying the history of medicine; offices and administration; a cafeteria; and student lounge. It is a complex and large building at 246,000 square feet (27,311 square meters). Care has been taken to segregate the circulation of students, staff, and biological matter, but without resorting to labyrinthine or windowless corridors. Its composition of contrasting atmospheres is theatrical, its use of materials masterful. Yet, its conceptual origins are remarkably simple. Gilles Saucier claims that his first sketch for the project (figure 8.3a and figure 8.3b) was "inspired by tree roots excavated from the building site"[1] and that the building's design attempts to find "a way to represent two trees interlacing like the roots coming down to the ground."[2] The University of British Columbia (UBC) campus was once a dense forest, giving cause for Saucier to elaborate: "There's an embedded memory of the site being covered with trees."[3] The tree metaphor is also linked by the architect to the history of pharmacology,[4] and, more broadly, it "serves as an allegory for the development of modern medicine."[5]

Saucier subscribes to a Romantic view of Nature as a source of spiritual renewal, declaring that on the weekends he goes "to the country and to my trees. They're waiting for me."[6] (We imagine the trees are less concerned with the weekend liaisons.) Nature is also for Saucier a muse and an ideal, a repository of form and gesture. Saucier draws, too, on contemporary formulations of ecology and environmental sustainability. Occasionally, nature figures as an actual tree to be preserved on site by judicious

Figure 8.2
The larger of the two atria of the Pharmaceutical Sciences Building (2012) at the University of British Columbia, Vancouver, Canada. Architects: Saucier and Perrotte. Photograph by Marc Cramer.

Figures 8.3a and 8.3b

Conceptual sketches for the Pharmaceutical Sciences Building (2009) by Gilles Saucier. Courtesy of Saucier and Perrotte Architects.

planning. It is not unusual for architects of Saucier's and Perrotte's generation to revere nature and to see themselves as working in harmony with natural forms and processes. Surprisingly, some of the discipline's most literate architects have led this charge, writing as Juhani Pallasmaa does that a "walk through a forest is invigorating and healing" and that "architecture is essentially an extension of nature."[7] Typically, as David Gissen has diagnosed, it is a benign and picturesque nature that is evoked—one that encompasses the most attractive and hospitable plants and landscapes (from a human point of view) and excludes the threatening, but equally natural, elements of decay and mortality, predation, violent weather, and natural disasters.[8] The forests Pallasmaa and Saucier frequent (in memory or reality) are not places of peril and exile as they were before the Romantic period. Nor are they like the ones we know in Australia (ours are menaced by leeches, mosquitoes, untethered flies, snakes, crocodiles, falling branches, and serial killers). The current enthusiasm for biological processes and biomimesis, which we introduced in the previous chapter, has only encouraged the idealized view of nature held by architects of a phenomenological persuasion.[9] What may at first glance appear to be divergent perspectives have a common belief that nature is honest, true, and intrinsically good. Nature is never "red in tooth and claw." Rather, a functional model of efficiency and judicious processes underpins an aesthetic model of strength and harmony. Metaphors drawn from nature serve both functional and phenomenological goals. Either way, there is no dearth of tree metaphors.[10] Sarah

Goldhagen attributes the popularity of trees to their being static, stable, familiar, and reassuring. People like trees, she claims.[11] The attraction for architects lies in their regular structure and irregularity at the fine grain, a contrast that is conducive to the design experimentation architects enjoy.[12]

Several laboratories are said to refer to tree forms and foliage patterns. At the Bryant Bannister Tree-Ring Building (2012), in Tucson, Arizona, a façade of articulated metal tubes is meant to resemble the leaves of the palo verde tree, indigenous to the Sonoran deserts of the southern part of the state. It seems appropriate. Scientists here research dendrochronology, the information contained in growth rings that develop over the life of the tree. Designed by Richärd and Bauer Architecture, the offices and laboratories are suspended above an open area of expressed diagonal supports that are often likened to a tree house. At the Institut National de la Recherche Agronomique (INRA) Research Laboratories (2013), in Charmes-la-Côte, France, by Tectoniques Architectes, the extensive use of timber is said to represent the research into forest genomics conducted in the laboratories. The façade "appears like a series of strips of timber on a landscape background."[13] Neil Durbach, of Durbach Block Jaggers architects, describes the façade design of the University of Technology Sydney's Faculty of Science and Graduate School of Health building (2015) as "inspired by a grove of rippling trees" (figure 8.4).[14] Wulf Architekten wrapped the Deutsches Zentrum für Neurodegenerative Erkrankungen (DZNE) (2017) with a façade of colored glass fins, "which draws attention to the forest, reflects it inward as well as outward, and takes on the colors of its foliage as they change with the seasons."[15]

Saucier and Perrotte also cite the tree as a model for the Anne-Marie Edward Science Building at John Abbott College in Ste-Anne-de-Bellevue, Quebec, completed in 2012, the same year as the PSB. It, too, has a foyer that twists and turns and intersects with a top-lit vertical circulation space. The architects refer to the space as the interior tree. Here, "an architectonic tree, analogous to that of the adjacent gingko," "contains the grand staircase and branches that extend through the building as built-in wayfinding elements and benches.... The vertical link thus becomes a public interior garden."[16] But Saucier and Perrotte do not only refer to trees. The architects are also fond of likening their signature use of fragmented planes and cave-like spaces to crystals, mineral strata, and geological plates. A 2013 multiresidential development for Toronto's West Don Lands is "composed of the elements found on the site: angular crystalline minerals of black and white.... The tower conceptually erodes to symbolize this dual mineral nature: a solid black object inset with white diaphanous crystals."[17] Their 2013 competition entry for the Pavilion 5 du Musée des Beaux Arts de Montreal is a "series of mineral strata,"[18] as is an indoor soccer stadium (2015), their second collaboration

Figure 8.4
The "rippling" façade of the University of Technology Sydney's Faculty of Science and Graduate School of Health building (2015). Architect: Durbach Block Jaggers. Photograph by authors.

with Hughes Condon Marler Architects, with whom they partnered on the UBC project. This complex is purported to grow from a park in Montreal as "a layer of mineral stratum recalling the geological nature of the site" and to be "like a series of luminous crystals among the trees."[19] Saucier credits his rural childhood and his previous studies in biology and botany with informing the firm's approach, claiming the Kamouraska, the region in Quebec where he grew up, to be "the source of his perception of an architecture inspired by geology."[20]

Although the architects describe the experience of the PSB foyer as evoking "a stroll through the space between huge trees or rock formations,"[21] it is the one instance where they refer to geological forms specifically in relation to this project, despite the compelling isomorphism (figure 8.5). Only the tree metaphor has been taken up and ratified by clients and commentators, with *Wallpaper magazine* referring to the building as a "kind of cubist tree." Similarly, *Architectural Record* embraces the laboratories as "tree houses," with offices and meeting rooms "like branches and foliage."[22] Eva Bjerring, writing for *Arcspace*, observes that the architects have "worked with the concept of

Figure 8.5
The cave-like main entrance foyer of the Pharmaceutical Sciences Building (2012) at the University of British Columbia, Vancouver, Canada. Architects: Saucier and Perrotte. Photograph by Marc Cramer.

the tree, where the branches intertwine creating a complete system of interconnections above the earthbound and life-giving ground level. The trunks … manifest themselves architectonically as the light filled atria that service all ranges of programs."[23] Mark Hume, writing for the local newspaper, *The Globe and Mail*, concurs with Saucier that the stairs draw "you naturally along, like a trail in an unknown forest."[24]

The resemblance to trees, though oft repeated, is not obvious. As metaphor, visual resemblance is not, theoretically, required. Metaphors describe one thing in terms of another to express commonalities not always apparent. For example, we speak of a close friend as "a rock" without necessarily imputing a physical resemblance. Where a building is isomorphic with that which it signifies, as in the brain-shaped International Neuroscience Institute (2000) in Hanover, Germany,[25] it is no longer metaphorical. It is a version of another thing in built form and on a different scale, as we explored previously in analysis of the Blizard Building. Such architectural gigantism raises the specter of the roadside duck-shaped stall, from which a vendor hawks eggs and fowl—the stall that was affectionately lampooned by Robert Venturi, Steven Izenour, and Denise Scott Brown in *Learning from Las Vegas: The Forgotten Symbolism of Architectural Form* (1977).[26] Because of the discipline's anxiety around literalness, the design process becomes one of generating, extrapolating, and then often hastily concealing one's metaphors. Alternatively, as theorist Charles Jencks recommends, the architect can multiply the "enigmatic signifiers to overcome the bane of the one-liner."[27]

Saucier expresses this anxiety when he confesses, "The difficulty of this project was knowing when to stop referring to the transformation of trees, because, of course, in the end it's about the spaces."[28] He labors to distinguish his use of metaphor from other science buildings in Vancouver wherein images drawn from science are displayed on their surface, citing the British Columbia Cancer Research Centre with its DNA-strand window patterns (it also sports a helical stair) as an example of what he believes is an ill-advised strategy.[29] The superior value of the use of metaphor in the PSB, Saucier submits, lies in its role as a three-dimensional generator within the whole of the design. The process has seen "the idea of a tree, whose branch system creates a canopy floating above the ground level" abstracted such that "it is subsequently given tectonic manifestation, and the architecture takes on a more geometric form."[30] Apart from Jean-Jacques Lequeu's cow, discussed in chapter 7, it is difficult to find architects who do *not* transform their sources into tectonic form—the advantage of scientific diagrams such as the DNA-strand over the tree is that they have already been subject to transformation and are famously "science-y." However, Saucier makes another, more unusual formal move (figure 8.6b). His tree concept becomes, in his words, "an inside-out" generator.[31] The idea is translated simply into the descriptor "interior tree."

Given the architects' use of their signature geotectonic forms alongside an allegedly arboreal spatial organization, the image of a cast or a rock revealing the imprint of a once-present plant is advanced. What is also suggested is that this tree is less tectonic and more stereotomic. That is, it is conceptually carved out of a solid rather than assembled as a frame, something hinted at in that early sketch (figure 8.3a and 8.3b). Goldhagen speculates on tree metaphors in contemporary architecture, which "simultaneously lament nature's absence and symbolically insert its presence."[32] In the case of the PSB, the tree's disappearance is made present as an absence. It is a double-negative sharing more, perhaps, with artist Rachel Whiteread's cast sculptures of since-demolished buildings, such as House (1993), than with the patently tree-like structures used by their architect contemporaries. As with Whiteread's work, a cast of something no longer present conveys a sense of loss and time's passing. Given the interiority and

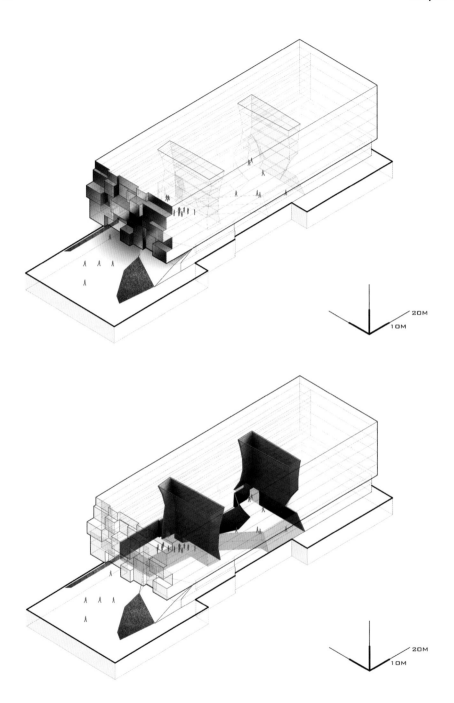

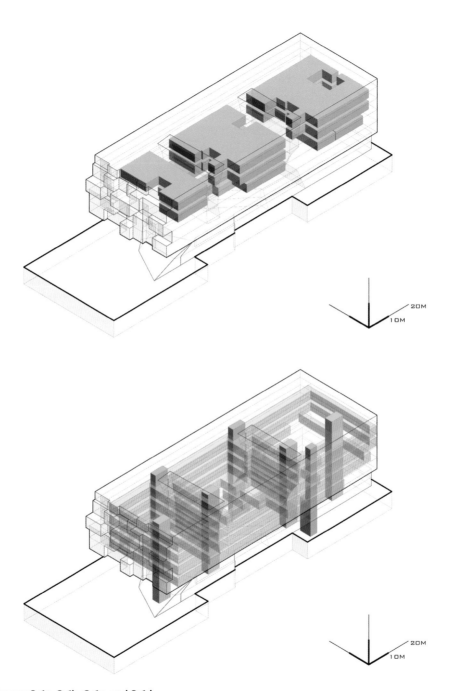

Figures 8.6a, 8.6b, 8.6c, and 8.6d
Diagrams of the architectural components of the Pharmaceutical Sciences Building (2012) at the University of British Columbia, Vancouver, Canada, showing (a) exterior, (b) the "inverted tree" voids, (c) laboratories, and (d) circulation. Architects: Saucier and Perrotte.

inversion of the tree form within a geologically inspired matrix at the PSB, it is an idea that suggests the fossil.

There are other reasons, though, to pursue analogies between this building and fossils and fossilization as a metaphor of laboratories more generally. Writing toward the end of the nineteenth century, Gottfried Semper contended that "old monuments are correctly designated as fossil shells of extinct organisms of society."[33] Laboratories experience this temporal dislocation more rapidly than most buildings, for science and its instrumentation are subject to breathtakingly fast cycles of novelty and obsolescence. We were often shown rooms constructed at great expense for equipment that had been superseded before laboratories were even opened. More critically, the fossil registers the transformation of a living organism into the nonliving. The process of fossilization is analogous to that effected in biosciences laboratories where living matter is rendered into code, text, and image. Both involve a kind of *writing* of the event of the organism's demise. Fossils and laboratories consume one type of life within another that is radically different from it. The most radical being expressed, or captured, is related to temporality. The fossil bears the traces of a prehuman event into the realm of thought. It relates to a time that precedes. It belongs to multiple lineages and disparate histories, which are constructed through it. It brings geological time scales to bear on and in anthropomorphic time. The temporality of the contemporary laboratory is radically different. We previously explored the idea that the focus of biosciences research tends toward the future, but it's a particularly anthropocentric future. Indeed, the story of the future told by most laboratories is not merely about the future of the species or the planet, but much more highly personalized. Its investments are framed in narratives concerned with your future, your parents' future and, most poignantly, your children's future. The coincidence of the temporalities of the laboratory and fossils makes their temporal co-presence both complex and perplexing.

Fossils in the Life Sciences

Fossils were crucial to the development of the life sciences, forming an archive increasingly revealed as industrialization opened the earth to quarry stone, mine coal, and extract precious metals and to build dams and other infrastructure. The paleobotanical record registers the historical evolution of plants and animals and the shifting geology of the earth. As Walter Alvarez succinctly states, "Rocks are the key to Earth history, because solids remember but liquids and gases forget."[34]

Before fossils were regarded as a record of once-living things made stone, they were a particularly curious puzzle. Because they register organisms no longer in existence, their unfamiliar forms confused early paleontologists. From a modern standpoint, the explanations given for their frequent and strange resemblance to human body parts and organs are curious. More unthinkable now is the magical power ascribed to them. In the seventeenth century, a medical colleague of the Danish natural philosopher Ole Worm recommended women wear around their neck "hysteroliths," the fossil of the brachiopod that resembles female genitalia, to loosen their constricted wombs.[35] Galileo's compatriot Frederico Cesi thought fossils were the partial transformation of earth and clays into plants. His botanical classification, published in 1651, arranged plants in a rising series from those he considered most like minerals to those most like animals.[36] Cesi located fossils midway between the mineral and vegetable kingdoms.

Debates raged throughout the seventeenth and eighteenth centuries—between Gottfried Wilhelm Leibniz, Nicolaus Steno, John Woodward, John Ray, Comte de Buffon, and others—as to whether fossils were the remains or residue of organisms or they were the inorganic products of the mineral realm that somehow mimicked the forms of organisms. Up until the posthumous publication of Leibniz's *Protogea* (1749), particularly troubling natural artifacts, such as crystals, were thought to belong with fossils. At the core of the debate was a question about the co-extensivity of human and geological history. If fossils are the result of altered sediments, then this would point to life consistently changing over a very long period—an idea contradicting Creationist doctrine. If fossils were fashioned by the power of geologic forces, as crystals and stalactites are, then it followed that the earth might be young and unchanged. Fossils could be the products of the original creation. In the context of this uncertainty, it was possible in 1726 for an established collector such as Johann Beringer to be defrauded by fake fossils procured and placed by two mischievous colleagues—this was despite the fossils boasting features never seen before, such as a scene of a comet with tail, a human face, and even Hebrew letters. Beringer recognized that these stones departed from other fossils, but dismissed claims they were carved by human hands. In part, as Stephen Jay Gould writes, this was because Beringer was arrogant, gullible, and craving the fame a unique discovery promised. The irresolution of debates about what fossils represented was also a factor.[37]

By the end of the eighteenth century, when a variety of fossils could be found across differing geological layers, it was accepted that fossils had been sequentially deposited and were of organic origin. Several scientists chronicled the successive assemblages of species and their extinctions to arrive at theories of pattern. In 1796, Georges Cuvier declared, "The theory of the earth is due to fossils alone," for "without them we would perhaps never have dreamt that there have been successive epochs."[38] Cuvier demonstrated for the first time the fact of extinction using small fossil fragments from which extinct animals could be reconstructed by a sequence of deductions based on the interdependence of each organic part. Cuvier did not, however, accept that species changed; indeed, the vast majority of eminent paleontologists of his time maintained the immutability of species. Confronted with the discovery of fossils of marine animals on land, Cuvier did, however, believe in catastrophic revolutions in which whole groups of organisms were replaced.

Cuvier's predecessor, the eighteenth-century French naturalist Jean-Baptiste Lamarck, proposed that all species found in the fossil record had become transmuted into modern forms, making extinction an impossibility—species simply changed into other species by mechanical means. For example, by stretching its neck to reach higher branches, the giraffe had slowly acquired its long neck. Darwin's insight, as we know, was to locate change not in individual organisms but in reproduction across a species. Postulating a single ancestor, Darwin was able to propose that all "species are the modified descendants of other species."[39] In 1859, when Darwin's *On the Origin of Species* was first published, the fossil record was poorly known, and he worried the lack of transitional fossils was "the most obvious and gravest objection which can be urged against my theory."[40] Despite the limitations of the available collections, he found sufficient patterns to support his theory of descent with modification through natural selection.

Darwin's debt to geology and paleontology was not just to what he learned from fossils. It was from geology and paleontology that Darwin inherited his historicism, for evolutionary biology is historical science. Life, he could see, was historical in the sense that the living organism was a record of the achievements of its ancestral line. Yet, science itself has a history, and the fossil belongs both to deep geological time and the human history into which it emerges as a talisman, sign of God (or his absence), a collectable, or a moment in an orderly sequence.

Fossils out of Time

If fossils have been significant in science, they have equally fascinated philosophers and historians. Michel Foucault makes the fossil a key figure in *The Order of Things: An Archaeology of the Human Sciences* (1966), his history of modes of scientific knowing and epistemic transformations. In the chapter "Monsters and Fossils" Foucault puts the fossil forward as the intelligible archive that emerges in contrast to the undifferentiated "endless murmur of nature."[41] The fossil, he argues, is the privileged object of resemblance for the historian of the continuum of evolution: "The fossil is what permits resemblances to subsist through all the deviations traversed by nature; it functions as a distant and approximate form of identity."[42] In the fossil, nature does not appear as a unified substance that contains and propagates itself. The fossil is a haunting figure or remnant from "that uncertain frontier region where one does not know whether one ought to speak of life or not."[43] Fossils come to us from a time before human existence, bearing witness to what we cannot know. They are, as Foucault writes, "a speech after death," which prefigures our own extinction as a species.[44] Foucault argues that nature is always understood through the lens of human knowledge, and knowledge is historical and a dense transfer point of power. Paleontology, thus, also has a history that brings concepts, methods, tools, and scientists to bear upon the fossil. The discovery of a fossil and its identification, circulation, and exhibition within museums or private collections add a new layer of human biography to the Earth, which accrues to the artifact itself.[45]

Bruno Latour finds a very clear demonstration of the coincidence of multiple, incompatible histories in two arrangements of horse fossils in the Natural History Museum in New York.[46] One conforms to an earlier belief in the linear evolution of the horse from small to larger, three toes to one. The other exhibit illustrates a twentieth-century understanding of the evolution of the horse with a more complicated and branching lineage, with smaller horses sometimes coming after larger ones. Latour describes the effect as contrasting "the progressive transformation of horses, and the progressive transformation of our interpretations of their transformations."[47] It is an exemplary illustration, he finds, of the double historicity of science and of its subject matter. The objects of science themselves have a history, and, conversely, science is formed by and bears its own historicity. The example supports his concept of the *factish*. A hybrid of fact and fetish, the factish is neither an independent reality that comes to be discovered after a successful scientific experiment, nor is it merely the projection of human beliefs onto an inert object. A factish involves both human and nonhuman actors. The successful scientific experiment is an example of a factish, for as Latour claims, it "is an event which offers slightly more than its inputs … no one, and nothing at all, is in command, not even an anonymous field of force."[48]

Latour is not arguing that the biological exists only as a human construct, and he rejects any anthropocentric account of nature. The challenge for philosophers of science, Latour tells us, is that "we have trouble focusing on the two aspects with equal emphasis: either we insist too much on the messy, mundane, human, practical, contingent aspects, or too much on the final, extramundane, nonhuman, necessary, irrefutable elements."[49] There is a breach between the horse, once alive, and its "life" as a set of bones collected, exhibited, and viewed in a museum's fossil collection. It is not, Latour insists, that being a bone carried along in paleontological networks is less rich or interesting than being a horse roaming free; it is that both belong to different vectors or histories. Much as land is reclaimed from the sea, Latour concludes, "knowledge is added to the world; it does not suck things into representations or, alternatively, disappear in the object it knows. It is added to the landscape."[50] He cites John Dewey, who in 1925 seemed to anticipate current debates. Dewey expresses the relationship between knowledge and science as something like a work of art. It "confers upon things traits and potentialities which did not previously belong to them.... Knowledge is not a distortion or a perversion which confers upon its subject-matter traits which do not belong to it, but is an act which confers upon non-cognitive materials traits which did not belong to it."[51] How can we think through the mutual implication of discursive practices and material phenomena in a way that in the encounter between culture and nature, each is active and transformative? Can nature do more than passively await Saucier's visits? This idea is beautifully illustrated by Roger Caillois's meditations on stones.

The Painted Stone

In *The Writing of Stones* (1970), Roger Caillois contemplates the allure of images perceived by imaginative observers of rocks and fossils that have been split open and polished. The resemblance between the patterns revealed by such a rock and, say, the tableaux of a burning town or a rampant dragon are improbable and arbitrary. Their appearance in the human realm is miraculous, "a fortuitous parallel—the result of some obscure physical and chemical reactions—between the patterns in the stone and a vague vision of half-ruined walls and towers."[52] They are also natural, existing undiscovered and outside of human knowledge until revealed by chance. In 1777, in a book entitled *Recueil des monuments des catastrophes que le globe terrestre a essuyées* (Collected vestiges of the catastrophes suffered by the terrestrial globe) the appearance of miniature cities and the like in rocks was conclusively dismissed by its author, George Wolfgang Knorr, as accidents or "Sports of Nature," which should, henceforth, be regarded as nothing more than a curious amusement. Previously, collectors probed their origins and meaning, eagerly seeking out and exhibiting those rocks that promised the greatest resemblance to pictures made by man. Once images are perceived, Caillois writes, "they become tyrannical and deliver more than they promised. The observer is always finding fresh details to round out the supposed analogy."[53] This is the problem of analogy that we spoke of earlier in chapters 6 and 7 and is amply demonstrated by the proliferation of the tree metaphor in architectural critique of the PSB. Indeed, fresh details were literally added by artists in the sixteenth and seventeenth centuries to complete the pictures they saw in rocks—for example, supplementing the buildings suggested by the stone with little black rectangles to indicate doors and windows.[54] In doing so,

Figure 8.7
Portrait of Roger Caillois in 1978. Photograph by Roland Minnaert.

the artist implicitly admitted nature could produce compositions that were art—in essence the painted rock is a co-authored work of art. Caillois's interest was not just in the rocks for themselves, which he avidly collected (figure 8.7). He was captivated by the coincidences across time and space that see meaningful correspondences between haphazard events and the logic of a human imagination that sees meaning in them. He declined to see meaning as simply imposed on the rocks, claiming instead that "the object makes a sign, becomes sign. It attracts onto itself that exact imagination, which reveals the object more than inventing it."[55]

If we were to place the PSB's tree metaphor in architectural history, we would likely begin with Marc-Antoine Laugier's Primitive Hut. Alternatively, we might trace the building's geological forms through a lineage that would start a little later, in the writings of the three seminal theorists of the nineteenth-century architecture, Gottfried Semper, Eugène Emmanuel Viollet-le-Duc, and John Ruskin.[56] From here we would go, perhaps, to Hans Poelzig's interior of the Grosses Schauspielhaus (1919) in Berlin, modeled after a cave with its stalactites. We might compare the PSB with the other examples of geological forms in buildings for scientific research—I. M. Pei's Mesa Laboratory for the National Center for Atmospheric Research (1967) in Boulder, Colorado, or, more recently, the Phoenix Biomedical Sciences Building (2017) in Arizona by CO Architects and Ayers Saint Gross. What such a history does not tell us, and what Callois would insist upon, is the material effect or agency rocks and trees have in their participation in architecture as matter and metaphor.

There is in the tree, perhaps, an anticipation of its architectural use, an invitation even. We see this invitation in the almost-already-architectural form of the trees in Charles-Dominique-Joseph Eisen's famous frontispiece for Laugier's *Essai* (figure 8.8). This anticipation of human desire is an opportunity for some species to dominate certain landscapes and to be transplanted around the world well beyond their original environment. This is an argument Michael Pollen makes in his discussion of the tulip, apple, potato, and cannabis plants in *The Botany of Desire: A Plant's Eye View of the World* (2001). Pollen observes in the cannabis plant's psychotropic appeal to the primate brain and in the tulip's beauty "a coevolutionary relationship" that has seen both flourish as a species under human cultivation. The Western red cedar (*Thuja plicata*), for example, which is used extensively in the PSB, was introduced from British Columbia where it is the provincial emblem, to western Europe, Australia, and the eastern United States. It is valued for its natural resistance to decay and insects and its light weight, straight grain, and tensile strength. These qualities give the tree longevity, but other qualities are of value primarily in a human context. For example, the perfumed aroma and warm sound of its timber have made it popular for guitars, bee hives, and saunas.

Unlike cedars and tulips, rocks are not impelled by mortality and the forces of survival of the fittest to trigger human desire, but do so anyway. In the beauty of rocks an alternative to the idea of the economy, or self-interestedness of nature, can be found. Caillois speaks of an "autonomous aesthetic force in nature."[57] In the extravagance of nature's forms and practices he finds evidence that nature is not a miser. For him, the extraneous beauty of stones tells of a tendency for superfluous splendor that nature shares with human art. The expressivity of geology operates as an expression not *of* something but expression that is *in and of itself*. There was geology before geologists. The fossil exercises a disembodied, nonhuman aesthetics.

We, along with other living organisms such as lichen and bacteria, are agents in monumental geologic processes. In the current geological age, referred to without irony

Figure 8.8
Charles-Dominique-Joseph Eisen, Primitive Hut, from Marc-Antoine Laugier, *Essay for Architecture* (1755), 2nd ed. Courtesy of the Royal Institute of British Architects (RIBA).

as "the Anthropocene," the fossil remains a harbinger of a future without humans. In this sense, the metaphor of a fossil for the architecture of a branch of the life sciences is brought into a new poignancy. The fossil blurs the distinctions between content and expression, the organic and inorganic. Through its ambiguous reference to trees and rocks, the PSB not only participates in what theorists have identified as a "geologic turn"—a preoccupation with rocks, earthquakes, and plate tectonics. It engages with current challenges to the categorical divisions between plants, animals, humans, and minerals.[58] Daniel Chamovitz, for example, believes that on the basis of new research humans must now see "that on a broad level we share biology not only with chimps

Figure 8.9
The tall and narrow "fissure" space between the offices and the laboratories of the Pharmaceutical Sciences Building (2012) at the University of British Columbia, Vancouver, Canada. Photograph by Marc Cramer.

and dogs but also with begonias and sequoias."[59] Others, such as Elizabeth Ellsworth and Jamie Kruse, go further, arguing that "it is becoming difficult for geologists and biologists to hold categorical distinctions between the 'brute materiality' of geology's 'external world' (rocks, minerals, mountains) and the soft, 'inner' worlds of biology's living things."[60] We now recognize that living beings are composed of geologic materials, such as calcium, iron, and phosphorous. Chemically, we are more like trees, rocks, and buildings than we care to admit.

The image of the absent tree, ossified within its geological matrix tension, captures the tension between the project of extending human life through pharmaceutical intervention and the sustainability agenda to which the PSB subscribes and for which it received Gold in the Leadership in Energy and Environmental (LEED) certification program.[61] Medical discoveries that extend aging populations in the resource-greedy First World are hastening an anthropogenic apocalypse. Here, the university's earnest attempts to remediate or repair the environment through its building and landscaping projects, and its research in ecology, are brought into contradiction with the pharmaceutical research program. In capturing this tension, the PSB building contains within it a passing resemblance to a fossil and whispers at its eventual submergence into the Earth's crust. It seems to be becoming-geological, already subject to the earth's tectonic forces.

There is a metaphorical appearance of minerals and trees in the PSB. This is supported by the actual extraction and refinement of minerals and timber evident in the building: the structural steel frame; the processed silica and limestone of the glass; and the sawn and finished timber that lines walls. In this sense, all works of architecture enact the coupling of matter and concept, nature and the technical. They are like Caillois's painted rocks, with some rocks especially suggestive of other worlds. PSB is such a rock. It is arguable that laboratories, by their very purpose, make a unique space for the observation of organisms from different temporal and spatial scales, indeed, from nonhuman worlds. Yet, it is the PSB that most intensely gives architectural expression to this quality. It does so in ways that are far more rewarding than could be anticipated by Saucier's early sketch of tangled trees. Through spatial negation and tectonic allusion, temporal contrast between historic pharmaceutical artefacts and contemporary visual technologies, the blurring of tree and geological metaphors, and the orthogonal container's denial of the diagonal internal rift, the PSB deftly brings into play the human and the nonhuman.

9   Floating

The previous three chapters in this section have focused on the analogical forms of architectural expression that proliferate in bioscience laboratories—the cells, mushrooms, origami, camels, polar bears, trees, and fossils that have found their way onto shimmering façades, or morphed into sculptural stairs and shapely voids. Through the translation and transformation of the objects of science these laboratories have attempted to popularize science, at the risk of betraying the seriousness of their biomedical endeavors. They have celebrated nature's forms, while the science has mined nature as a ready resource. We have arrived at an architectural exuberance that "makes things special" in an epoch of rampant exceptionalism.[1]

This chapter considers the near impossibility of the translation of the ideas and ideals of experimental biomedical science into architecture, and particularly, how one laboratory—Charles Correa's Champalimaud Centre for the Unknown (2010) in Lisbon, Portugal—is explicitly dealing with this impasse (figure 9.1). We suggest that Correa makes of this laboratory an "empty" or "floating" signifier. This is an idea developed in the anthropology of Marcel Mauss and Claude Lévi-Strauss to describe a word, event, or gesture that is void of meaning and thus apt to receive any meaning: a signifier without a signified. This notion belongs also to the discourse of Gilles Deleuze and Félix Guattari's *a-signifying* semiotics. For Deleuze and Guattari there is no sanctified relation between signifier and signified (that is, meaning). This is described as the *univocity* of the sign relation. This univocity plays itself out in the formulations of delirium, stutter, and exhaustion, which Deleuze celebrated in the *Critical and Clinical* (1998) project.[2] It is in events of delirium, stutter, and exhaustion that one's ability to speak finds physical form and not coherence. Words are coughed out, and expression becomes more a matter of gesture than of meaning. These events occur in excess of fixed and simple meanings, metaphor, and analogy. Indeed, they occur beyond the logics of linguistics. We are suggesting for an architecture to *speak* of science it is likely that it, too, needs to operate in a similar realm, beyond signification.

Figure 9.1
Streetview of the Centre for the Unknown (2010), Lisbon, Portugal. Architects: Charles Correa. Photograph by authors.

In many respects the Centre for the Unknown partakes of "the atmospherization or mundanization of contents" that Deleuze and Guattari speak of in *A Thousand Plateaus* (1980).[3] The content of this laboratory and clinical facility is not elaborated or expressed in the architecture. The Centre for the Unknown consists of a biomedical laboratory and clinical treatment and education buildings for the Champalimaud Foundation. Correa does not offer an image for the brave new world of biomedicine, of which the foundation speaks. Nor does he establish an image for the workplace culture of its scientific community.[4] Instead, the architecture gestures toward a horizon that lies beyond the known contents of science and the hoped-for discoveries of its scientists. As the expressive force of architecture gestures elsewhere, the contents of science become vague. In this regard, the architectural expression is disentangled from the ideas and ideals of the biosciences, while being responsive to the creative sense of possibility that the sciences hold. That is, expression becomes disentangled from content. This chapter considers how the architecture of biosciences research might operate as what Deleuze and Guattari describe as an "amorphous continuum that for the moment plays the role of the 'signified.'"[5]

The Centre for the Unknown

At the 2010 inauguration of the Champalimaud Centre for the Unknown, Correa proposed:

> What makes me most proud about this project is that it is NOT a Museum of Modern Art. On the contrary, it uses the highest levels of contemporary science and medicine to help people grappling with real problems: cancer, brain damage, going blind. And to house these cutting-edge activities, we tried to create a piece of architecture. Architecture as Sculpture. Architecture as Beauty. Beauty as therapy.[6]

In subsequent publications of Correa's speech, the phrase "real problems" was underlined, pointing to the disjunction that occurs in architectural thinking between content and expression—as if the latter were less real or important than the former. It also points to an anxiety that architecture merely houses, or decorates, the reality of the world. Correa's deferral to real problems is an odd linguistic formulation given the abstract intentions and less-than concrete program of the Centre for the Unknown.

According to one of the directors, João Botelho, "We call it the Centre for the Unknown, because, likewise, our discoveries are from the realm of the unknown."[7] In fact, the unknown element of the Champalimaud Foundation's mission begins with the vagueness of the bequest established by the Portuguese industrialist and financier, António de Sommer Champalimaud. The former Portuguese Minister of health, Leonor Beleza, alongside Champalimaud's long-time friend and lawyer, Daniel Proença de Carvalho, were responsible for constructing the foundation's lofty ambitions, if not its intended activities.[8] The "first challenge was to identify the areas in which the endowment … could be put to the best use" and "to find the areas of intervention most aligned to what António Champalimaud himself would have wanted."[9] It was the "would have wanted" that was open to discussion. After several months the foundation settled on oncological and neuroscience research, soon after adding blindness prevention. International researchers and specialists in translational medicine in these fields were to be brought together, in a newly built center, with clinical specialists. The "translational" movement of experimental science through to clinical trial and application is one of the key aims of the Centre for the Unknown. Botelho suggests, "The building is not neutral. It is part of this project."[10] We have seen such assertions of agency before in descriptions of the Janelia Research Campus.

The Centre for the Unknown is located on the coast of Pedrouços, to the north of Lisbon, on the northern bank facing the River Tejo, close to the point at which it joins the Atlantic Ocean. The Centre covers a site area of over 646,000 square feet (60,000 square meters) and consists of three main structures. The first, and largest, is the Champalimaud Clinical Centre, a four-story structure on the north side of the site. The building contains treatment and clinical facilities on lower levels and laboratories on upper levels. From the building lobby, patients visiting the facility can get glimpses of the research laboratories above. The building can accommodate 440 scientists. Glass walls on all levels overlook a large verdant garden that is accessible only to staff and those receiving treatment after entering the building (figure 9.2a and figure 9.2b). This 27,000-square-foot (2,500-square-meter) three-story high garden is full of tropical plants from Africa, India, Brazil, and Timor. In an interview about the building, Correa noted "the healing presence of rain forests. In other words, Nature herself."[11] There are many odd inversions involved. The exterior plaza beyond this space is stark, dry, and white, while the inside is green and humid. The only tree in the plaza outside is an ancient olive transplanted from the interior of Portugal, while inside is all manner of imported plants. This space is like a greenhouse within a walled garden. It has a pergola roof that repeats the orthogonal glazed pattern of the laboratory and clinic walls. This garden thus occupies an interstitial space, between the orthogonal regularity of the main building with its efficiently planned interiors and the sweeping curve of the *exterior* wall and plaza beyond. Correa repeats this pattern of engaging gardens to negotiate the gap between the utilitarian spaces of science and medicine and the expressive exterior. A second, smaller garden located at the east end of the tear-drop built form also functions in this way. This garden, referred to as the "Zen garden," is open to the sky and accessible only to the patients of the clinical center.

The orthogonal interior spaces and gardens are embraced by a sweeping four-story wall. In plan, the wall appears as a large portion of an ellipse drawn into a tear-drop shape. This rounded, smooth, white stone wall is perforated with large ovoid (egg-shaped) openings. Like the portals of cathedrals, the scale of these openings is proportioned not to the human body, but to the wall in which it is placed. It has little to do with the demarcation of inside or outside, for both sides of the opening are, for all intents and purposes, outside—the courtyard and Zen gardens on one side, and the plaza on the other, are both open to the sky.

A second structure, toward the south of the site, houses the auditorium, restaurant, and the exhibition area on the entrance level and a conference center and the Champalimaud Foundation administration offices above. This structure also has an elliptical, or tear-drop, floor plan and is connected to the research laboratories at an upper level by a glass tube bridge 69 feet (21 meters) in length. The auditorium has a large ovoid window with views to the river and the Belém Tower in the distance. The impressive sixteenth-century Manueline tower is associated with the Age of Discovery and the period of Portuguese global imperialism. The exhibition space on the ground floor of the Centre for the Unknown hosts events related to the foundation's scientific endeavors. From the restaurant ("Darwin's Café") a generous terrace extends toward the water. Here, we were served cool mojitos by an elegant but harried waiter overwhelmed by the Friday-night crowd. Unlike other on-site cafes and bars that largely serve the laboratory occupants, Darwin's is a destination.

A third structure is as much landscape as building. It is more open than that which we tend to refer to as "a building" and more closed-in than that which we tend to refer to as "a landscape." It is best referred to as "a plaza" (figure 9.3). The plaza of the Centre for the Unknown glides between its two main buildings as the plaza of the Salk Institute for Biological Studies does and, like it, is empty of vegetation and people. The Champalimaud homepage declares the creation of spaces "with the public in mind" to be "enjoyed by all," but adds the caveat: "The public use of these spaces is subject to the decision of the management of the Champalimaud Foundation which can at any moment alter the rules of use."[12] This idea of a public space that is nevertheless privately managed is one we've seen before at the Salk Institute. In other ways, however, the Centre for the Unknown plaza departs radically from Louis Kahn's example. The sky above the Salk Institute's plaza, and the view to the sea it frames, are defined by perspectival order and a sense of geometric precision. In contrast, the Centre for the Unknown's plaza is viscous, asymmetrical, and curvaceous. We noted earlier Paul Goldberger's description of the Salk Institute's plaza as being "set in magnificent juxtaposition to the infinite openness of the ocean."[13] By contrast, views to the water at the Centre for the Unknown are oblique and deferred. The usual hierarchies of empty space and object, inside and outside, which are so firm in the Salk Institute, have been rejected. We look from an open plaza through ovoid openings up *into* a garden. It seems it is the plaza that generates the sweeping white walls enclosing the structures to the north and south, rather than the other way around. The plaza conists of an open-air amphitheater and an open public landscape. The semicircular stone amphitheater hosts concerts and public performances with the river as a backdrop. Counterintuitively, the plaza slopes upward at a rake of 1:20 toward two monoliths and the river. Correa writes:

> At the end of the ramp are two stone monoliths, straight from the quarry, as primordial as Stonehenge. When you reach the highest point, you begin to see a large body of water–which seemingly connects to the ocean beyond. In the center of the water body, just below the surface of the water, is an oval shaped sculpture—made of stainless steel and slightly convex, so that it reflects the blue sky and the passing of the clouds above.[14]

The stainless steel, spherical body in the water looks as though it is the back of a huge turtle, an emerging planet, or a suspended air bubble expelled by a docile sea monster.

Figures 9.2a and 9.2b
Views of the interior garden between the laboratory and clinical wing of the Centre for the Unknown (2010), Lisbon, Portugal. Architects: Charles Correa. Photographs by authors.

Figure 9.3
The plaza of the Centre for the Unknown at twilight (2010) showing the elliptical opening in the main research building. Architect: Charles Correa. Photograph by José Carlos Sousa.

Correa suggests of this "enigmatic object": "It could be an island it could be a Portuguese man-of-war."[15]

The Floating Signifier

For Claude Lévi-Strauss the "highly esoteric" ethnological and sociological work of Marcel Mauss provides an opportunity to reassess how meaning operates.[16] Lévi-Strauss's *Introduction to the Work of Marcel Mauss* (1950), published the year of the sociologist's death, reflected a time when anthropology was operating as a key field between the sciences and the humanities. Mauss had conceded to the idea that social phenomena were linguistically constructed. Yet his work led him to cultural products and events that were socially operable while falling short (or beyond) the logic of linguistics itself. Mauss identified "a representation which is singularly ambiguous and quite outside our adult European understanding," which he variously called an "embodied elemental force," "magical thinking," or "mana."[17] It is this idea that led Lévi-Strauss to the notion of the "floating signifier":

> I believe that notions of the *mana* type, however diverse they may be, and viewed in terms of their most general function (which, as we have seen, has not vanished from our mentality and our form of society) represent nothing more or less than that *floating signifier* which is

Figure 9.4
The plaza of the Centre for the Unknown (2010), Lisbon, Portugal. Architect: Charles Correa.
Photograph by authors.

the disability of all finite thought (but also the surety of all art, all poetry, every mythic and aesthetic invention), even though scientific knowledge is capable, if not of staunching it, at least of controlling it partially. Moreover, magical thinking offers other, different methods of channelling and containment, with different results, and all these methods can very well co exist. In other words, accepting the inspiration of Mauss's precept that all social phenomena can be assimilated to language, I see in *mana*, *wakan*, *orenda*, and other notions of the same type, the conscious expression of a *semantic function*, whose role is to enable symbolic thinking to operate despite the contradiction inherent in it.[18]

The work challenges the fundamentals of Ferdinand de Saussure's linguistics.[19] For Saussure, though the relationship between the signifier and signified is arbitrary, it is also fundamental. That is, one is never present without the other. The implication of a floating signifier is that any "primacy of the signifier," as Jacques Lacan later refers to it, is questionable.[20] In this sense, the submerged spherical body at the end of the plaza is not a turtle, a planet, an air bubble, an island, *or* a man-of-war; it is a turtle, a planet, an air bubble, an island, *and* a man-of-war and so on and so forth. There is no need for any one sanctified meaning to dominate when the object operates as a floating signifier. This spherical body signifies—but what it signifies remains "singularly ambiguous." The disconnection between signifiers and signifieds means that signs might be operative in their own right. That is, a sign can be operative without being connected to any fixed meaning.

Roland Barthes writes of the *mana* of architecture when he turns to the Eiffel Tower in *Mythologies* (1957). He notes the moment when a young Gustav Eiffel, aged twelve, "discovered the 'magic' of Paris" while on a visit from Dijon with his mother.[21] The Eiffel Tower itself would come to operate with the magic of the empty signifier. Barthes writes:

> This pure—virtually empty—sign—is ineluctable, *because it means everything*. In order to negate the Eiffel Tower (though the temptation to do so is rare, for this symbol offends nothing in us), you must, like Maupassant, get up on it and, so to speak, identify yourself with it. Like man himself, who is the only one not to know his own glance, the Tower is the only blind point of the total optical system of which it is the center and Paris the circumference. But in this movement which seems to limit it, the Tower acquires a new power: an object when we look at it, it becomes a lookout in its turn when we visit it, and now constitutes as an object, simultaneously extended and collected beneath it, that Paris which just now was looking at it.[22]

Barthes echoes Lévi-Strauss's language in referring to a "'floating chain' of signifieds" and notes what Lévi-Strauss referred to as "the inherent contradiction in it."[23] It is this mobility of sign relations that leads Barthes to introduce a sliding scale of *iconic*, *motivated*, and *arbitrary* signs. The iconic tends toward one function—that is, one singular link in meaning. The crucifix in Christianity, for example: it might be elevated in a

Gothic cathedral or hang low on the hirsute chest of a Portuguese waiter, but its meaning has not altered. The motivated sign, on the other hand, has a mobility to it. That is, the sign might stand in place of multiple meanings. The double helix as announced in *Nature* (1953) takes on a different, yet, connected meaning when installed as a staircase in the headquarters of a pharmaceutical company.[24] For Barthes, the arbitrary sign is one that is near empty of fixed meanings; instead, the meaning of the sign is entirely given by the contingencies of its use. We noted Spikey at the Blizard Building has a relationship to a viral cell only because of its context; were it placed elsewhere it would be something radically different. Barthes counsels that "every code is at once arbitrary and rational."[25] The architect of the Centre for the Unknown is aware of the mobility of meaning. Correa refers to something like Barthes's notion of the *iconic* as the "primordial symbolism" that architectural spaces, open to the sky, can incite.[26] Correa also observes the arbitrary nature of some signs, noting "the mud pot, used in an Egyptian village to draw water from the well, has completely different connotations when caught in a beam of halogen light at the Metropolitan Museum of Art."[27]

Beyond Barthes's sliding scale, Lévi-Strauss's notion of the floating signifier suggests there need not be any connection between a sign and meaning. Deleuze and Guattari succinctly summarize Lévi-Strauss for their own purposes, stating that "the world begins to signify before anyone knows *what* it signifies; the signified is given without being known."[28] In chapter 8, we explored the expressions of geology as belonging to prehuman eras. The banded stratifications of geology, the casts of fossils, and the shimmer of crystals were expressive well before anyone attempted to extract meaning from or apply meaning to these formations. These geological signs were produced without deferral to meaning. Here, there was production before what we recognize as product. In *A Thousand Plateaus* signs fall into machinic systems of production, in which meaning is but one of many outcomes. The philosopher and psychoanalyst identify multiple "regimes of signs" through which meaning is constructed. Such regimes are operative because of the mobility of the sign and the manner by which it "floats." Deleuze and Guattari write:

> All signs are signs of signs. The question is not yet what a given sign signifies but to which other signs it refers, or to which signs add themselves to it to form a network without beginning or end that projects its shadow onto an amorphous atmospheric continuum. It is this amorphous continuum that for the moment plays the role of the "signified," but it continually glides beneath the signifier, for which it serves only as a medium or wall: the specific forms of all contents dissolve in it.[29]

In *Essays Critical and Clinical* (1993) Deleuze contends that medicine (the clinical) might distil from art, and specifically literature (the critical), a symptomatology.

Symptomatology is concerned not with the play between signifiers and signifieds, nor with origination, nor with "giving a reason," nor with causation (as with etiology). Symptomatology is the study of signs—full stop.[30] Symptomatology as such may be thought of as a logic of floating signifiers. Daniel W. Smith, a Deleuze scholar and one of the translators of the text, suggests that "while etiology and therapeutics are integral parts of medicine, symptomatology appeals to a kind of limit-point, premedical or submedical, that belongs as much to art as to medicine."[31] Such a position reminds one of Lévi-Strauss's suggestion that the floating signifier is "the surety of all art, all poetry, every mythic and aesthetic invention."[32] Deleuze turns specifically to literature (the critical) as a means of investigating this medical (clinical) notion. He suggests writers do not account for or necessarily represent the world, but rather construct or compose worlds from indeterminate symptoms. Geologists are involved in a similar process, collating data and images in a manner that translates stone into stories of the birth of the Earth, generating the factish. Writers compose the terrains, interactions, inversions—the sense of alternate worlds that constitute modes of existence. There is in symptomatology an enfolding of contexts and selves, and a constructing and configuring of both. For Deleuze, "The ultimate aim of literature is to set free, in the delirium, this creation of a health or this invention of a people, that is, a possibility of life."[33]

The architecture of the Centre for the Unknown is involved in a similar enterprise of health and the possibilities of life, not just in the sense that it houses the contents of biomedical science, but, rather, because of its expressive force. This proposition can be explored in three ways: with respect to its *inverted interiority*, its *dislocations* from reason, and the profound *emptiness and exhaustion* that the Centre for the Unknown enfolds.

Inverted Interiority

The entrance of the main building on the northern side of the Centre for the Unknown is clearly marked, but visitors tend to be drawn past it, through the plaza space and upward toward the water. The plaza organizes the site at the same time that it disorganizes it. It is the heart of the project, but has itself no center. Correa makes similar inversions in his multiple unbuilt works, such as the farmhouse for Indira Gandhi (1972) and the Kapur Guesthouse (1976), as well as the Museum of Archaeology in Bhopal (1985). Such projects place "the highest emphasis" on what Correa refers to as "open-to-sky space."[34] The patterning of courtyard spaces in relation to interiors, Correa calls "the Inside-Out Sock."[35] Correa's sock operates a little like Gilles Saucier's "interior tree," inverting a more traditional logic to approach the design of a space afresh. That key dualism between interior and exterior that characterized Correa's

Figure 9.5
The white plaza of the Centre for the Unknown (2010), Lisbon, Portugal. Architect: Charles Correa. Photograph by Peter Westerhof.

architecture is compromised somewhat at the Centre for the Unknown. The open plaza operates "inside-out" in multiple respects, but the sense of a space being either one *or* the other is lost. The near-white Portuguese Lioz limestone cladding of the exterior of the Centre for the Unknown creates complex relations that empty, or hollow-out, the site (figure 9.5). On a summer day, the white walls and pavers underfoot produce a scintillating, near-blinding, glare—as if simulating the experience of partial blindness studied within. At the same time, we are drawn to the plaza, and although it operates as point of desire, it is not homely in any sense. Indeed, it is discomforting and foreign, empty and raw. It is hard to see even through squinting eyes.

The glare operates much as Deleuze describes light operating in German expressionistic cinema. In *Cinema 1: The Movement-Image* (1983) the philosopher describes a "light which has become opaque, *lumen opacatum*," and suggests that "from this point of view natural substances and artificial creations, candelabras and trees, turbines and sun are no longer different."[36] Glare is the blinding and disorienting materiality of light. It is able to express and, at the same time, obscure the content of sight. Glare is a materialization of light. The glare of the plaza space produces a *zone of indiscernibility*—where the clear and geometric distinctions of form and plane are lost. The glare reflecting off the rounded walls, the water, and the plaza paving is not the glare of dialectical opposition—a glare between dark and light, or between matter and the immaterial; rather, it is a glare with a material presence of its own. Lévi–Strauss refers to the "apparently insoluble antinomies"[37] involved in *mana*. The glare generated by the Centre for the Unknown produces something similar: it dissolves walls, floors, water, and sky. Glare operates as an expressive *quality* and what it obfuscates is *quantities*. The measureless and formlessness of glare plays itself against the most measured of quantities here: the geometries of the architecture and the precisions of biomedical research.

Dislocations

The site for the Centre for the Unknown is organized around a diagonal line that marks the course of the central plaza space. Correa writes of the first time he saw the site, "I knew that whatever else we did, the site must be structured along a powerful architectural diagonal axis, an open-to-the-sky space, going right from the entrance to the opposite corner, where you finally see the river beginning to merge with the ocean and the great unknown."[38] The line is, in fact, a diagonal arc that tends to be followed in a cinematic sweep until it reveals the river and the sea beyond. The logic here is not of the orthogonality of laboratory organization, access, egress, and the physical containment necessitated by biomedical research. The organization of the floor plans makes it very clear as to the dislocations of content and expression in this building. The laboratories and clinical spaces are organized on a grid of columns and linear circulation. The internal planning is not unusual for a biomedical research laboratory and clinic. The emphasis is on efficiency and a form of organization one might associate with accountancy and economy. The internal gardens (the sunken garden, the infusion garden and the terrace garden) mediate between the refined efficiencies of the internal layout and the equally refined, but geometrically unlike, expressions of the exterior architecture (figure 9.2a and 9.2b). If one were to remove the curved walls, the building would almost certainly still stand and its function would be almost unaffected. However, it would be a radically different prospect and place. It would put into play a radically different logic.

Correa writes of the "*Genius Loci*, the essential meaning of a site," and suggests it is "architecture's unique responsibility to express to release, that meaning."[39] We counter, however, that meaning is not the point at the Centre for the Unknown. The genius loci is the protective spirit of a place drawn from Roman religion. However, Correa is referring to the version of the idea that relates to a "spirit of place" as the Norwegian architect and phenomenologist Christian Norberg-Schulz was to reformulate it, based on a sentimental and selective reading of Martin Heidegger's "Building, Dwelling, Thinking" (1951).[40] For Norberg-Schulz, every place has a particular meaning that an architect must responsibly articulate or amplify. Norberg-Schulz imagines, "Architectural history shows that man's primeval experience of everything as a 'Thou,' also determined his relation to buildings and artefacts. Like natural elements, they were imbued with life, they had *mana*, or magical power."[41] The architect in such a formation is a *shaman*, channeling the forces of the earth, a soothsayer of "essential form" and meaning.[42] Correa refers to "Architecture as a Model of the Cosmos—each expressing a transcendental reality, beyond the pragmatic requirements of the programme that caused them to be built."[43] The architecture of the Centre for the Unknown is the generation of a particularized cosmos, but not one that grounds us as Norberg-Schulz supposed. It is an alien landscape gesturing to the cosmos, but disassociates itself from the historic context of Lisbon's urban fabric. The generation of difference between Correa's buildings and their settings is often written about in terms of the architect's penchant for the generation of microclimates. However, there is something particularly singular to the Centre that is far more expressive than it is functional, but no less operative.

In writing of the Eiffel Tower, Barthes insists that "architecture is always dream and function, expression of a utopia and instrument of a convenience."[44] Barthes tenders it was

> as if the function of art were to reveal the profound uselessness of objects, just so the Tower, almost immediately disengaged from the scientific considerations which had authorized its birth (it matters very little here that the Tower should be in fact useful), has arisen from a great human dream in which movable and infinite meanings are mingled: it has reconquered the basic uselessness which makes it live in men's imagination. At first, it was sought—so paradoxical is the notion of an empty monument—to make it into a "temple of Science"; but this is only a metaphor; as a matter of fact, the Tower is nothing, it achieves a kind of zero degree of the monument; it participates in no rite, in no cult, not even in Art; you cannot visit the Tower as a museum: there is nothing to see inside the Tower.[45]

Emptiness and Exhaustion

It is this "nothing to see" that makes the Centre for the Unknown so enticing. In engaging with the unknown, the architecture becomes a point of departure and a gesture "to infinity."[46] Like the empty central space of the Salk Institute, it engages what Correa refers to as the "metaphysical aspects of the sky."[47] But it does not connect the contemplative or collective work of science with that open-endedness. In Correa's essay "The Blessings of the Sky," the architect notes the singular and central value of the sky in "human history," "since the beginning of time."[48] The sky becomes very much a key part of this architecture. Correa declares that the sky operates as a meta-phor "for our relationship to something outside (and beyond) ourselves."[49] The over-arching sense one has here is of emptiness, both metaphorical and real. The contrast between the lively space of the restaurant terrace, filled with people, mojitos, deck chairs, and graphic images of scientific motifs, and the empty plaza could not be more pronounced. It is a pared-back, open emptiness. The impotence of the architecture makes the sensation more potent.

In *Essays Critical and Clinical*, Deleuze suggests that "the sign that refers to other signs is struck with a strange impotence and uncertainty, but mighty is the signifier that constitutes the chain."[50] The odd alien minimalism of the plaza and the buildings that flank it generates the most intense sensations. The space makes no attempt at a totalizable, or photogenic, image for itself. The architect-pilgrim does not know where to stand to capture the iconic photograph: only moving images could convey its cin-ematic qualities. This architecture is mobile and doesn't rest in *this or that* metaphor or *this or that* symbol. Its material is less the discernible stone and curve than the indis-cernible sky and glare. For Correa,

> We live in a world of manifest phenomena. Yet, since the beginning of time, man has intui-tively sensed the existence of another world: a non-manifest world whose presence underlies—and makes endurable—the one he experiences every day. The principle vehicles through which we explore and communicate our notions of this non-manifest world are religion, philosophy and the arts. Like these, architecture too is generated by mythic beliefs, expressing the presence of a reality more profound that the manifest world in which we exist.[51]

The two monolithic concrete stelai (for they are in fact not stone), each 50 feet (15 meters) tall, which mark the step from land to river, themselves bleed into the sky (figure 9.6). Correa proposes that the columns "announce the presence of the Infinite Unknown that lies beyond."[52] The ends of the columns have been painted blue. It sounds like a rather simplistic idea—to paint the end of a column blue and imagine that it might bleed into the sky. But on this Friday, in this bright sun, the treatment of the columns lets them leave the earth entirely. There is a reflecting pond beyond the columns, and beyond that the river and the sea. One element slides into another pro-ducing what Deleuze and Guattari refer to as an "amorphous continuum that for the moment plays the role of the 'signified'."[53]

This zone of indiscernibility between the architecture and landscape, water and sky is highly productive. The minimalist treatment of the architecture is removed from the operations of the biomedical endeavor—and yet not entirely. It is refraction rather than reflection. They breed fruit flies at the Centre for the Unknown because of the ease by which they can be genetically manipulated. They also breed shoals of zebrafish

Figure 9.6
The concrete columns at the Centre for the Unknown (2010), Lisbon, Portugal. Architects: Charles
Correa. Photograph by authors.

because of their uncanny ability to regenerate organs without scarring. The architecture is similarly malleable and generative, open to possibility and the contingencies of the unknown as it is open to the sky. One can imagine that Lévi-Strauss was speaking of the Centre for the Unknown when he wrote of *mana* that it is

> Force and action; quality and state; substantive, adjective and verb all at once; abstract and concrete; omnipresent and localised. And, indeed, *mana* is all those things together; but is that not precisely because it is none of those things, but a simple form, or to be more accurate, a symbol in its pure state, therefore liable to take on any symbolic content whatever? In the system of symbols which makes up any cosmology, it would just be a *zero symbolic value*, that is, a sign marking the necessity of a supplementary symbolic content over and above that which the signified already contains, which can be any value at all, provided it is still part of the available reserve, and is not already, as the phonologists say, a term in a set.[54]

Part III: Socialization

The following three chapters elaborate an idea that has quietly but consistently occupied earlier chapters. This is the simple idea that the laboratories of bioscience generate scientists as much as they generate science. In the background of many of our investigations has been the scientist. We have witnessed the anthropological construction of the scientist in the work of Bruno Latour and Steve Woolgar; the scientist of the voyeuristic gaze observed behind vitrine walls; the introverted scientist of the island; the bickering scientist in a red space-suit who is the subject of media focus; and the scientist of mission statements, contracts and wi-fi-enabled shuttle-buses. We now focus more explicitly on the ways by which laboratories construct these scientists, and in turn the ways in which scientists construct themselves through architecture.

We have used the word "socialization" to refer to the mechanisms by which the laboratory constructs the scientist, but other words are equally relevant here. Michel Foucault would call it "subjectivation." Maurizio Lazzarato would speak of "subjective processes," and Gilles Deleuze and Félix Guattari refer to the impersonal and pre-individual senses of "desiring-production." We chose to stick with "socialization." This word emphasizes the construction of subjectivity that occurs under broader social and disciplinary gazes, and it speaks of the pressures upon scientists to literally socialize as much as it does of the collective and interdisciplinary ambitions of contemporary bioscience. While the scientist has her own point of focus (a subject or product of her own), she too is made the subject of the biosciences. She is subjected to and internalizes containments and controls. Her ambitions are harnessed and her desires directed. The laboratory operates as a rich and elaborate network of relations. The scientist is but one of its products.

10 Symbiosis

On a bleak October day at the Cold Spring Harbor Laboratory (CSHL) on the North Shore of Long Island, New York, we took refuge from the rain and pondered the intersections between science, philanthropy, and architecture. The gazebo in which we sheltered is adorned with a copper filial modeled on an adenovirus, but our attention was drawn to the names of places and buildings that we had seen on our tours of the campus.[1] The monthly public Sunday tour had started at the Oliver and Lorraine Grace Auditorium. We had slept in the austere single rooms of the Charles and Helen Dolon Hall and woken to the bells ringing in the Lita Annenberg Hazen Tower. We had strolled up de Forest Drive, past Tiffany House and as far as the Mary D. Lindsay Child Care Center. We had seen Nicholls Biondi Hall, the Carnegie Building, and the John Divine Jones Laboratory. Each building's name acknowledged the significance of its patrons. There were also named rooms, such as the Plimpton Seminar Room and the Howard Hughes Teaching Laboratories in the Arnold and Mabel Beckman Laboratory. Equally, we noted the memorialization of living and dead scientists—the Demerec Laboratory, the Delbruck Page Laboratory, the Luria Cabin, the Watson Crick Bridge, the Barbara McClintock Laboratory (formerly the Animal House), and others.

It was the unsettlingly nostalgic architecture of the most recent addition to the campus—the Hillside Campus (2010), designed by Centerbrook Architects—that had brought us to the CSHL. Here each of the structures that rises above its six subterranean laboratories also make clear its philanthropic debts (figure 10.1). The research at the Nancy and Frederick DeMattheis laboratory focuses on the genetic basis of diseases, including cancer, autism, and schizophrenia. The Donald Everett Axinn Laboratory is for research on the neurobiological roots of mental illness. The David H. Koch Laboratory houses an interdisciplinary team who develop mathematical approaches to interpret and understand complex biological data sets. The William L. and Marjorie A. Matheson Laboratory undertakes research on the tumor microenvironment

Figure 10.1
Aerial view of the Hillside Campus courtyard and upper level entry (2010) at the Cold Spring
Harbor Laboratory, Long Island, New York. Architect: Centerbrook Architects. Courtesy of CSHL.

and metastasis, and the Leslie and Jean Quick Laboratory develops new therapeutic strategies for treating cancer. Lastly, the Wendt Family Laboratory is for research on neurodevelopment.

Naming rights, be they for gazebos, buildings, professorial chairs, or prizes and scholarships, register philanthropic interests in ways that promise the name is spoken again and again, as in a refrain. As we noted in chapter 2 with regard to the Claudia Cohen Hall at the University of Pennsylvania, building names are sold and resold. The naming of places is a form of politics, a way of creating connections with the past and influencing memory and culture in the present. For the cultural geographer Derek Alderman, place-naming is a means "of claiming the landscape, materially and symbolically, and using its power to privilege one world view over another."[2] In the case of Long Island, it is not insignificant that the Native American name for this place was *Paumanok*, meaning "The Island that Pays Tribute"; for the original inhabitants paid a form of potlatch or tribute to the more powerful tribes surrounding them, just as building names do.

Philanthropy at the Cold Spring Harbor Laboratory

Commemorative naming proliferates at the CSHL in large part because of its Public Charity Status as a private, not-for-profit organization.[3] In 2017, the laboratory had an income of USD 213 million, only one-seventh of which was drawn from federal government grants.[4] Thus, most of its research funding comes from public support, as well as investments and royalties, and its research activities reflect the concerns of individuals and philanthropic foundations. Over its 120-year history the CSHL's research program has mirrored the changing interests of the state and of benefactors, and they will change again. In the 1950s, its scientists used a grant from Schenley, the whisky distillers, to work on improving antibiotic yield. During the same decade, Harold Abramson was studying psychotropic drugs and testing LSD on Siamese fighting fish and carp. He was also participating as a subject in experiments on the effects of LSD on humans, which were carried out at Mt. Sinai Hospital and funded by the CIA. Today, one team researches the in vitro identification of candidate genes for disorders such as autism—a focus aligned with the interest of board member and donor Marilyn Simons whose daughter was diagnosed with autism at the age of six.[5] Nothing, however, marks the intersection between the interests of donors and science more sharply than the CSHL's infamous role as the center of eugenics in North America.

Between 1904 and 1941 researchers at CSHL advocated for the improvement of the human species by "discouraging" reproduction by persons having genetic defects (negative eugenics) coupled with encouraging reproduction by persons presumed to have desirable inheritable traits (positive eugenics).[6] Eugenics studies were extremely successful in attracting donations and students. In 1910, Charles Benedict Davenport founded the Eugenics Records Office (ERO) of the Carnegie Institution's Station for Experimental Evolution, with sponsorship from Mary Williamson Harriman, widow of a railway magnate.[7] Harriman purchased a mid-Victorian timber residence nearby on 75 acres (30 hectares) for the ERO and paid for a new masonry wing. The ERO soon enrolled far greater numbers of students than other courses.[8] Harriman subsequently gifted a new brick building in the Second Renaissance Revival style and USD 300,000 to enable the Carnegie Institution to endow a Department of Genetics at Cold Spring Harbor. The department advised state governments on sterilization legislation.[9] We wondered, in the gazebo, with Harriman's legacy just out of sight (but not mind), how scientists and philanthropists currently negotiate their individual agendas.

Contemplating the interactions between wealthy individuals, scientists, and the establishment of shared territory, be it architectural or intellectual, we considered Darwin's analogy of the "tangled bank."[10] The tangled bank paragraph is one of Darwin's rare textual flourishes, wherein he describes the complex relationships between species and place. As the harbor darkened before us, we recalled the debates around social Darwinism, as it has been promoted by figures such as Ernst Haeckel, Francis Galton, and Herbert Spencer. Too often, the "laws of nature" have been extrapolated to human affairs to excuse sexism, racism, and violence as immutable behaviors inherited from our ancestors or related to our survival. We were, of course, at one of the places in which social Darwinism had taken a dark turn, with Davenport ignoring the impact of the environment altogether in pursuit of a pure genetic determinism (he even proposed that a love of the sea ran in families). But it is Darwin's emphasis on competition and predation as one of the primary impetuses in natural selection that is the most pressing limit for understanding altruistic behavior and cooperation.

It was not until 1879, in the publication of Anton de Bary's *Die Erscheinung der Symbiose* (The Phenomenon of Symbiosis) that mutually beneficial relationships between organisms were named.[11] The definition de Bary gave to *symbiosis* referred more broadly to any association between species, but biologists now use this term to refer only to persistent alliances where all participating organisms benefit. Organisms have repeatedly responded to antagonists, such as predators, and abiotic stresses, such as low nutrient availability, by forming mutually beneficial alliances with other organisms, which result in the enhanced fitness and ecological success of all the participants. These relationships are not straightforward and symmetrical. They might vary seasonally or circumstantially and involve an exchange of very different functional capabilities—for example, defense for food or clean teeth for protection. In many ways the characteristics of symbiotic relationships resemble the regularized but asymmetrical interactions that are the key feature of Peter Galison's concept of the trading zones of science. Galison explains:

> I hand you a salt shaker and in exchange you pass to me a statuette.… We do not in any sense have to agree to the ultimate use, signification, or even further exchange value of the objects given.… While for me the statuette may be a religious object, for you it could be a purely aesthetic or functional one.… We strip away meaning and memory when we pass the object to a trading zone.[12]

Just as social Darwinism pitted human against human, others have been equally keen to make analogies between symbiosis and human society, using one as an analogy for the other and vice versa. We detect ideological intent in the blurring of the lines between the biosciences and the social sciences by Richard Dawkins, Peter Kropotkin, and others. The debates range from attacks on altruism to desires to see, or make, human society more cooperative in line with those observed in interspecies relationships. Peter Corning, for instance, believes cooperation enhances the survival and reproduction of the human species.[13] He deploys scientific models to advocate for social justice and "the fair society." Lynne Margulis, one of the more forceful advocates for revising evolutionary theory in the light of symbiosis, uses examples from human society to explain nature. Margulis proposes, "The problem is not 'competition versus cooperation.' … Even bankers and sports teams have to cooperate to compete. When you compete … you still cooperate!"[14] The analogical upscaling of symbiosis might be useful in exploring the relationships between different people and organizations as they are carried out in place and under specific conditions. However, we would again note the cautions related to the different scales of analogy discussed in chapters 6 and 7.

We decided nevertheless to test the model of symbiosis on a specific interaction relevant to the CSHL, which had attracted the attention of the mainstream press. On December 4, 2014, the medal for the Nobel Prize in Psychology or Medicine awarded to James Watson in 1962 was auctioned at Christie's, New York.[15] Watson was jointly awarded the prize along with Francis Crick and Maurice Wilkins "for their discoveries concerning the molecular structure of nucleic acids and its significance for information transfer in living material."[16] Watson's association with the CSHL is a long one. He first visited in the summer of 1948 as a graduate student with his thesis advisor, Salvador Luria. In 1953, Watson and Crick made their first public presentation of the DNA double helix at the CSHL annual summer symposium, which that year was focused on "Viruses." In 1968, Watson married Elizabeth Lewis and became the CSHL's director. He was appointed president in 1994 and chancellor ten years later.

Figure 10.2
Mosaic of James Watson in the Jones Laboratory at Cold Spring Harbor Laboratory (1989), Long Island, New York. Photograph by Peter Menzel, courtesy Science Photo Library.

While on a book tour to the UK in 2007 Watson was quoted in the *Independent* as saying he was "inherently gloomy about the prospect of Africa [because] all our social policies are based on the fact that their intelligence is the same as ours," a comparison he doubted.[17] Seemingly echoing the ideas of the CSHL's eugenicists of a century earlier, he said, "People who have to deal with black employees find this not true."[18] The CSHL made a public show of disowning Watson's comments, demanding his apology and retraction, and suspending him as chancellor. Soon after, he claimed to be "short of money after being made a pariah"[19] and put his Nobel Medal up for auction. The medal sold for USD 4,757,000 to Russian oligarch and co-owner of the Arsenal football club, Alisher Usmanov. Usmanov promptly returned the medal to Watson, explaining that it was "unacceptable" that an outstanding scientist should be brought so low. Given that his own father had died of cancer and Watson's contributions in this field were invaluable, Usmanov wanted the scientist to donate the proceeds of the sale to research institutions that had supported his career.[20]

We had seen Watson driving his vintage green Jaguar towards his harbor front residence at the northern tip of the campus and were confident that his impoverishment and banishment were imagined. Indeed, he remained a valued employee continuing to guide the architectural development of the campus and engage in its philanthropic and scientific events. In 2016—years after his supposed expulsion—Watson's salary from the CSHL as its Chancellor Emeritus was USD 372,446, with an additional USD 175,020 in other compensation.[21] The CSHL's report to the Internal Revenue Office in 2016 claims that Watson is still working a forty-hour week. If the sale of the medal was not about money for Watson, it was even less so for Usmanov. With assets estimated by Forbes to be worth USD 15 billion, Usmanov's gesture came at negligible personal cost. Watson had given the Russian the opportunity to grandstand. In our opinion, the scientist had solicited public sympathy and ironically been given the opportunity to be a philanthropist. We do not know if Watson became a donor, or instead purchased the David Hockney painting he had intimated he might buy with the proceeds.[22] In 2019, Watson repeated his belief that blacks are genetically inferior to whites in intelligence, and the CSHL, again, distanced themselves from his remarks and revoked his title of Chancellor Emeritus.

Symbiotic interactions are now recognized as having significance for entire ecosystems—indeed, as being their constitutive feature. We know that even our alimentary canals are unique ecosystems containing large numbers of interdependent (tangled) microbial species. Michelle Speidel suggests, in relation to biological evolution, "Perhaps it would be better to see [symbiotic associations] not so much in terms of what each partner is getting out of the relationship, but how the structure as a whole is functioning."[23] This may be a useful concept that helps us to understand the exchange between Watson and Usmanov as something more than a farcical diversion. In doing so, we need to leave behind the view of symbiosis as borne out of insufficiency. Extending this idea, we can see that philanthropy does not arise because there are pockets of need. Rather, it arises out of an economic system in which a small minority of individuals accrue financial excess. It stems from having too much.

Earlier we explored how, for Georges Bataille, excess requires expression and release. It might also require legitimation. Getting rich for Slavoj Žižek is "a violent process of appropriation which casts doubt on the right of the rich giver to own what he then generously gives."[24] At the core of Žižek's disgust, and his battles with Peter Sloterdijk on the subject, is the notion that philanthropy perpetuates the conditions that cause inequality. Patricia Nickel also considers that "the pursuit of ostensible social change through genuine social exclusivity is one of the key practices through which governing takes place."[25] That is, philanthropy valorizes the wealthy and benevolent subject and thus addresses a deficit in governing at a time when inequality is pronounced. Nickel insists that the very visible practice of philanthropy through gala balls, named buildings, and other forms of publicity is a kind of "disinfectant" against possible opposition to wealth concentration and inequality.[26] In this view, philanthropy produces the demand for scientific research in order that excess might be redirected in ways that seem to be of noble benefit and public good. Such cynicism towards acts of altruism (albeit related to familial benefit) remind us of the conclusions of Dawkins's *The Selfish Gene* (1976), wherein the cellular process of genetic replication is anthropomorphized, while sentient beings are reduced to machines.

Our only disagreement with Žižek's misanthropic assessment of philanthropy is his focus on contemporary forms of charitable consumerism. His critique of the movement, hinged on Starbucks, TOMS shoes, and benefactors such as Bill and Melinda Gates, draws attention away from the social milieu in which philanthropic practices have been longstanding. Indeed, Žižek's assessment fails to show how the social capital of families with "permanent wealth" is accrued through philanthropy. Coming together on the boards of charities and gala fund-raising events is as important in defining cultural capital as attendance at polo matches and membership of sailing clubs. For CSHL, the philanthropy of local individuals has been foundational and persistent across its 120 years of operation. In 1993, Watson frankly described the situation in an interview while still director of the CSHL. As a manager of a scientific research institution, he claimed: "You have to like people who have money. I really like rich people."[27] Scientific research as a whole has established institutions, subdisciplines, processes, events, and projects that capture and repurpose excessive wealth. Architecture might be one such project.

Historic Entanglements

To understand how the Western red cedar shingles of the gazebo in which we are sitting (built in 1976 but mimicking an earlier style of architecture) participate in, and further, the philanthropic economy and networks of the CSHL we need to relay something of the history of the CSHL (figure 10.3). The CSHL grew from two co-located but operationally distinct institutions that merged in 1963: the Bio Lab and the Carnegie Institute. From the beginning, both institutions were conceived, funded, and, sometimes, managed by private donors. During the eighteenth century, grist and woolen mills, barrel factories, and shipyards occupied the site that is now the CSHL. Between 1836 and 1858 whaling became the main industry, and the town boomed. In 1859, the same year as the publication of Darwin's *On the Origins of Species*, the discovery of petroleum in Pennsylvania precipitated the end of the whaling industry. These unrelated events led to the establishment of a seaside zoological station in 1890.

Heirs of the Jones family whaling fortune had established the Wawepex Society to manage their real estate and to invest funds in scientific research (to the indigenous

Figure 10.3
View of the Cold Spring Harbor Laboratory on Long Island, New York, from across the harbor. Photograph by authors.

inhabitants, Cold Spring Harbor was known as Wawepex, or "at the good little water place," for its abundant spring of freshwater).[28] In 1888, the Wawepex Society leased land in the area to the Brooklyn Institute of Arts and Sciences for a fish hatchery and biological laboratory, which was clearly intended to rival the recently established Marine Biological Laboratory at Woods Hole, Massachusetts. John D. Jones, Bio Lab co-founder and president of the Wawepex Society, put up USD 5,000 to construct the first purpose-built laboratory there in 1893—a timber building in the Colonial Revival style. The Wawepex Society leased 10 acres of land for fifty years to the Carnegie Institution of Washington for a Station for Experimental Evolution from 1904, under the leadership of Charles Davenport. The tradition of building, or rebuilding, houses for new directors and their families began that year with a grand house for Davenport. In his report to the board of managers of the BioLab, Davenport emphasized the ways in which the new house made it possible to "pay some of the social debts that had accumulated."[29] It was in this same decade that New York's wealthy industrialists built their suburban mansions along Long Island's North Shore coast—a place and a period immortalized in F. Scott Fitzgerald's novel *The Great Gatsby* (1925).

At the turn of the twentieth century, its clambakes, bathing, and boating—along with its location among the weekend homes of New York's *best* families—made the Bio Lab an attractive destination for those who attended its summer research camps. Indeed, in 1899, Gertrude Stein, then a medical student, and her brother, Leo D. Stein, a PhD in biology, were among the attendees. Something of the life of the summer researchers there can be glimpsed in a complaint made by the neighbor to the north, Henry W. de Forest. In 1902 de Forest wrote to the Wawepex Society about the behavior of the laboratory students cavorting on his beachfront. Director Davenport defended them to Walter Jones in his reply: "Every one of the days of work is crowded full to the brim.... If the students go bathing on hot days at 5 o'clock it is because the bath enables them to work better.... I assure the Society that no loafer or mere pleasure seeker is admitted to this Laboratory if I know it."[30] Eventually, the curmudgeonly de Forest paid for five cedar boats so that the students could row to the small portion of the Sand Spit belonging to the Bio Lab without having to pass below his house. Here, the exchange of boats for privacy is clear.

The BioLab, which had continued independently of the Carnegie Institution, came under the control of the newly formed Long Island Biological Association (LIBA) in 1924. Its first president was investment banker Marshall Fields and its board members included luminaries of New York society, such as William K. Vanderbilt, Childs Frick, Louis Tiffany, and de Forest. Most of these LIBA directors had residences in the area. Well into the 1960s, a highlight of the annual symposia was when "speakers went to

the homes of LIBA members for dinner parties that brought them together with prominent figures in the local community."[31] In 1942, the spit of land, which the students
had once traversed to the dismay of de Forest, along with a carriage house and stables,
was given to the BioLab by his widow. Over the next decade, the BioLab expanded its
research program. While it had funds for projects and summer workshops and symposia flourished, the buildings and grounds had become run down and overcrowded.
Many buildings were unheated and limited to summer use. Most were domestic or
industrial structures unsuited to the work scientists wished to undertake. Enter Watson
and his penchant for social networking. Today the LIBA remains a nonprofit organization that represents the "friends of the Laboratory." LIBA is made up of twenty-five
trustees, most of whom are not involved in the day-to-day operations of the laboratory.
However, it appears that the personal histories of the LIBA trustees are well enmeshed
with the contemporary investments of the laboratory.

Watson's Architectural Patronage

When the Watsons arrived at the CSHL, the grounds were "disorderly and delightful
with cover, vibrant with birds and animals."[32] Their living accommodation, however,
was substandard. Having holidayed at the Sea Ranch in California, the Watsons quickly
appointed the architect of its lodge and condominiums, Charles Moore, to undertake
the renovation of Airslie, a timber farmhouse built in 1806. Watson recalls that "once
we moved in, we realized that Moore's unique design gave Liz and me a way of life
usually enjoyed only by the very wealthy."[33] The experience piqued Elizabeth Watson's interest in architecture and propelled her to undertake a master's degree at the
Columbia University School of Architecture and Planning in 1983. Her thesis became
the basis for *Houses for Science: A Pictorial History of Cold Spring Harbor Laboratory* (1991),
published by the CSHL on the occasion of its one-hundred-year anniversary. She subsequently wrote *Grounds for Knowledge: A Guide to Cold Spring Harbor Laboratory's Landscapes and Buildings* (2008). Both books are a curious mix of scholarship, pedantry,
affectionate anecdote (gossip), and public relations. Elizabeth Watson also drafted the
nomination papers that led to the placement of the laboratory's main campus on the
National Register of Historic Places in 1994, the same year that Centerbrook Architects
completed the Watsons' new home (figure 10.4).

For Watson, the experience with Moore, along with the tours of stately English
homes he made with Elizabeth, energized his efforts toward securing funds for the
improvement and expansion of the CSHL's built infrastructure. In *Avoid Boring People* (2007), Watson advises future managers of scientific institutions not to scrimp on

Figure 10.4
Ballybung, the Watson's house at CSHL (1994). Architects: Centerbrook. Photograph by the Brenizers.

building facilities with cheaper materials and unknown architects. This, he writes, "is bad for business in the long term…. Less than sturdy structures give out messages that their institution's life may be equally short. In contrast, solid, stylish buildings give donors the confidence that their descendants will one day bask in reflected glory. The lure of permanence inspires generosity."[34] And this may be the crux of the value of architecture to individuals with philanthropic means. Watson's astute understanding of the motivations of his philanthropists extends to his recognition of the need to emphasize the social continuity between them and the laboratory's scientists. He insists:

> Research institutions must have rich neighbors nearby who are inclined to take pride in local accomplishments…. Entering worlds where your trustees relax—joining their clubs or vacationing where they go with their families in the summer, for instance—is a good way to put relations on a social footing. Seeing you as more friend than supplicant will incline them to go the extra distance for you in a pinch.[35]

Watson was adept at playing both supplicant and friend as the situation demanded, and affected a certain *sprezzatura* as he donned the caricature of the socially awkward scientist. Before meetings with prospective donors, he was observed to untie his shoelaces and muss his hair.[36] In his tribute to Lita Hazen, one of the CSHL's late donors,

Watson relates that she "enjoyed bright and unusual people who would say things that she wouldn't expect."[37] We suspect he was describing himself.

While the location of the laboratory at Cold Spring Harbor is a historic accident, Watson was correct in predicting the continued benefits from being situated among New York's most expensive real estate. The area, despite Davenport's nightmare of a nation of mixed race people, is also one of the United States' whitest regions. The population is 97.03 percent white, compared with 65.7 percent for New York state.[38] Laurel Hollow, the location of the campus, has a population of less than 2,000. Once the home of Louis Comfort Tiffany, it is one of the wealthiest towns in America.[39] Based on an analysis by Little Rock's Gadberry Group, nine towns on Long Island's North Shore made *BusinessWeek*'s list of America's twenty-five wealthiest towns.[40] Members of CSHL's board of trustees have properties in these towns. Among its most generous donors—giving more than USD 5 million to the CSHL—three are on the board of trustees: Marilyn Simons, Jamie Nicholls, and Lindsay Goldberg.[41] Simons and her husband, Jim, made their fortune through a hedge fund company that uses his mathematical expertise and customized software to profitably invest its clients' funds. We have already noted their role in expanding research into autism and antenatal diagnosis.[42] Vice president of the board, Simons lives in Manhattan and at East Setauket, located farther east from CSHL. Jamie Nicholls, who was elected chair of the board in 2010, lives with her financier husband at Mill Neck, just three miles (five kilometers) from CSHL. Mill Neck was once home to the Vanderbilts, Whitneys, Rockefellers, and Levitts (the developer of Levittown). Making donations through their foundation, the King Street Charitable Trust, Nicholls and her husband, Fran Biondi, are amongst the CSHL's donors of more than USD 5 million. They have inherited wealth and Ivy League educations.[43] Nicholls's great grandfather, Joseph Treneer, invented Alka-Seltzer, a pain reliever launched in 1931.[44] Her late father, Richard Hall Nicholls, specialized in tax-exempt bonds and was fondly described in his obituary as "the tax lawyer's tax lawyer."[45] An "avid sailor, skier and swimmer,"[46] he died in 2009 after a ten-year struggle with an incurable form of cancer.[47] His daughter is also on the board of Memorial Sloan-Kettering Cancer Center. Her mother, Judy Cormier, has an "appointment-only" interior decorating business and is a regular attendee of the CSHL's Double Helix Awards dinners.[48] The Nicholls-Biondi Hall opened in August 2015, just months before our visit, but with its low vaulted-roof looks as though it dates from the 1980s (figure 10.5). Charles and Helen Dolan, who own Madison Square Garden and founded Cablevision and HBO, funded the dormitories at CSHL. They live nearby at Oyster Bay. Mary Lindsay, for whom the child care center is named, was a sister-in-law of John. V. Lindsay, a former mayor of New York City. Active in family planning and Planned Parenthood,

Figure 10.5
The Nicholls Biondi Hall at CSHL (2015). Architects: Centerbrook. Photograph by authors.

Mary Lindsay and her lawyer husband lived in Laurel Hollow. Donald Everett Axinn is an author, poet, aviator and developer of office and industrial parks throughout New York City. He lived on Long Island with his wife, Joan, who was a member of the Sands Point Country Club and the Old Westbury Racquet Club. His name adorns one of the buildings of the Hillside Campus.

Of course, other scientific institutions benefit from philanthropy and use this money to build new research centers.[49] What is unique about the CSHL is the depth and longevity of its relationships with the local community. Of the approximately fifty donors who gave more than USD 25,000 in 2014, only a handful were corporate entities with a potential interest in the results. CSHL's donors are often motivated by their personal experiences of illnesses that are the focus of the laboratory's research. Equally, they identify with the organization as one of their own, much as they might their country club. That is, the CSHL is more than a place associated with their philanthropic community—it is a microcosmic reflection of it. Much of this has to do with Watson. Speaking to schoolchildren in 1975, Watson declared his hope that science would one day not be limited to those bright children "whose enthusiasm for learning makes them congenitally unable to put up with the polite banter that characterizes the successful lawyer or banker or businessman."[50] Nevertheless, he has himself not only put up with lawyers, bankers, and businessmen, but cravenly sought them out. He has eschewed professional fund-raising for a more personal approach to "potentially generous neighbors."[51] The CSHL courts donors through an annual gala dinner (with tickets priced between USD 2,000 and 12,000) at which the Double Helix Medal is awarded to "individuals who have positively impacted human health by raising awareness and funds for biomedical research."[52] Board members Marilyn and Jim Simons, as well as David Koch, the billionaire supporter of conservative causes and opponent of the 2010 Affordable Care Act, are previous recipients of this medal. The event attracts celebrities, scientists, philanthropists and broader New York society. These include Craig Venter, the first to sequence the human genome and whose J. Craig Venter Institute in La Jolla, California, is the subject of the next chapter. Another attendee is the architect Richard Meier, who has designed several large houses dotting the Long Island coast in a mannered homage to the modernist Le Corbusier.[53]

Looking Back at the Hillside Research Campus

The CSHL, while producing scientific research, reproduces socially and architecturally a version of the residential villages around it.[54] Where Moore's early interventions on the campus were witty—his Sammis Hall (1981) is purportedly modelled on Palladio's Villa Poiana of 1549, albeit in stucco—those of his successor firm, Centerbrook Architects and Planners, denizens of a converted mill in Essex, Connecticut, have become increasingly eclectic and nostalgic. Their design for the Computational Neuroscience Laboratory recalls the "novelty siding of the original cabins" and is roofed in copper foil shingles.[55] Like many of Centerbrook's designs for CSHL, the shingles and classical shape of the gazebo in which we sat works hard to conceal its purpose and youth. The Beckman Laboratory (1981) tries to conceal its size as well as age through "its dark brick exterior" and "extra large windows that make it appear smaller when viewed from a great distance."[56] Elizabeth Watson optimistically proposes that "it could be mistaken for a grand water view–endowed Long Island mansion design in classical turn-of-the-century-style."[57] This is not so. It looms over its neighbors, and the architects have taken pains in subsequent projects to break up the volume of new buildings and orient their narrower elevations toward the harbor. Centerbrook's second home for the Watsons, this time a new build, is in the English Regency style, painted a peach color and featuring symmetry, chimneys, and traditional double hung windows. It has a separate underground two-car garage connected by a subterranean tunnel to the house (figure 10.4). Built in 1994, and pretentiously christened "Oaks at Ballybung," the house, according to Elizabeth Watson, is "suitable for large-scale entertaining." More pretentiously, it was "inspired by the classic farmhouses outside Venice designed in the late sixteenth century by the Italian architect and author Andrea Palladio."[58]

The first major expansion of the infrastructure of the CSHL took place in 2009 with the opening of the Hillside Laboratories at a construction cost of USD 100 million, 80 percent of which came from private donors and philanthropic foundations.[59] Housing about one-third of its research personnel, the new laboratories are below ground and have no natural light or outlook. They are also inflexible, dispersed, and awkwardly interconnected.[60] There is no flexibility in the planning that would allow research teams to occupy less or more laboratory benches as projects wax and wane. Indeed, the compromised functionality of these laboratories underscores the rhetorical priorities of the CSHL's architecture. Above ground six discrete buildings, vaguely resembling bloated versions of the structures that composed the original whaling village, are clustered around a multilevel courtyard (figure 10.1, figure 10.6). Each building is painted a different color—sienna, sage, olive, umber, yellow ochre—a technique to disguise

Figure 10.6
The courtyard of the Hillside Campus at Cold Spring Harbor Laboratory. Photograph by authors.

size.[61] The roofs are steeply pitched and punctuated at each end by vertical "chimneys" concealing the necessary vents and risers of the hidden laboratories. Randal Jones, the head of facilities management at CSHL, explains the design "was intended to recall an alpine village. This is enhanced by the severely sloping site, the use of artificial pavers in the courtyard spaces, and a towering central exhaust stack mimicking a church bell tower common to village squares."[62]

Over a period of thirty years, James Childress has been personally involved in fifteen architectural projects at CSHL. Childress describes the campus as "the antithesis of a large, boxy, factory laboratory setting bristling with characterless buildings, macadam, and white-coated drones." This deferral to an ahistorical caricature of science (and the scientist) is fascinating for an architect fixated with historical reference and nostalgic inclinations. Childress goes on to suggest of CSHL that "it is a Village for Science, with intimate spaces indoors and out so researchers can bump into one another, conspire, perambulate, gossip, bird-watch, or meditate. One might mistake Cold Spring Harbor Laboratory for a small, private day school."[63] In a series of negations, Bill Grover of Centerbrook suggests, "We didn't want to build something that would make it no longer look like a small whaling village."[64] It would seem to be a simpler task to state what the architecture of the CSHL is "no longer" or "is not" than to state what it is. Childress believes the buildings of CSHL "do not look new or even like laboratories."[65] He adds, "It's not obvious, even from close up, what goes on at Cold Spring Harbor Laboratory."[66] In 2009, a reviewer for the New York Times suggested that "an architectural sleight of hand has disguised the new labs as a miniature Bavarian hilltop village."[67]

Centerbrook call this architecture "American place-making," but it is a very narrow slice of America that is given place in such architecture.[68] Jorge Otero-Pailos retrospectively describes the Critical Regionalism that firms such as Centerbrook favor and continue to produce, as a withdrawal from, or sublimation of, the political in favor of the experiential and phenomenological.[69] In *The Architectural Uncanny* (1992), Anthony Vidler observes that "the postmodern appeal to roots, to tradition, to local and regional specificity" is driven by a "renewed search for domestic security."[70] Both see such architecture as overdetermined by political circumstances. It is not that the architects who design the vaguely "classicist" houses, country clubs and educational buildings on America's West and East Coasts are passive conduits. Indeed, many are vocal proponents of premodern styles. It must also be noted that the retrogressive architecture that we see at CSHL is not the new or discrete phenomenon suggested by architecture's neat historical categories. Between the 1890s and 1940s, the elite families of the region built over 1,000 mansions in neo-Georgian, English Tudor, Gothic, Roman, and French chateau styles (and combinations thereof) in emulation of the country estates of the

European aristocracy. They employed architects such as the Beaux-Arts educated William Delano, who designed in Cold Spring Harbor the French chateau styled Oheka for Otto Kahn in 1919. It is these houses now owned by the CSHL and its neighbors and friends that the new buildings on campus emulate.[71] Today's conservative elites employ Robert Stern Architects, Shope Reno Wharton, Haynes-Roberts, and others specializing in "traditional architecture with contemporary comforts." Centerbrook provide a similar service to the CSHL.

Instead of boasting their influence, Centerbrook repeatedly ascribe architectural decisions to James and Elizabeth Watson, claiming the couple have "guided our hands (slapping them occasionally)."[72] Centerbrook principal William Grover recalls on his first visit to the campus that hours were spent "divining the exact curve of a driveway that would satisfy Jim Watson's sense of balance and symmetry. We knew this wouldn't be easy, but we still had (and still have) great respect for his motivation."[73] With the architect's cooperation and expertise, the Watsons have overseen the development of the campus toward the creation of a pseudo-historic architectural ensemble that is idiosyncratic in the field of the biosciences. We have come across just one other contemporary laboratory building that mimics the pitched roof and gable, materiality, and style of a vernacular agricultural structure—named the Eva J. Pell Laboratory for Advanced Biological Studies at Pennsylvania State University. Its laboratory has a Biological Containment Level 3 classification because its scientists study diseases more infectious than its architecture, such as avian influenza, plague, and West Nile virus. The laboratory is situated on the outskirts of the university's campus, its barn-like design driven by the need to reassure and disguise its purpose. There is a little of this impetus in the architecture of the Cold Spring Harbor, but the detailing of more recent buildings is nostalgic in a way that the Pell Laboratory is not.

Svetlana Boym, writing on nostalgia, asserts:

> One is nostalgic not for the past the way it was, but for the past the way it could have been. It is this past perfect that one strives to realize in the future…. Nostalgic reconstructions are based on mimicry; the past is remade in the image of the present or a desired future, collective designs are made to resemble personal aspirations and vice versa.[74]

Her observations are most relevant here. The CSHL campus does not look the way it does just because of a preservation order. It is not preservation that is being practiced, or recreation, but the realization of an entirely new and idealized village. The CSHL has fetishized and cannibalized its architectural inheritance to the point where it has become impossible to distinguish between the old, the restored, the reproduced, and the newly built. Although Watson and his architects cooperated on a vision that at times seems driven by the scientist's memories of Cambridge and stately homes in England, the objective aims at a more general appeal to the North Shore milieu. It is not a degraded version of an original and authentic village, for there never was one like this. It is a graft, a new village of science hosted by its neighbors. Through the architecture of the CSHL a philanthropic base is constructed and reified, a scientific agenda forged and favored, and the excesses of the capitalist economy expressed and modulated in maintenance of the status quo. Here both scientists and philanthropists find succor.

11   Aggrandizement

This chapter is concerned with a building where the reputation and interests of the organization's lead scientist, J. Craig Venter, are paramount to its design. The three-story headquarters of the J. Craig Venter Institute (JCVI) in La Jolla, California, opened in 2013 and was designed by Zimmer Gunsul Frasca at a construction cost of USD 48 million (figure 11.1). It consists of a laboratory and administration facility of 45,000 square feet (4,180 square meters) on a 1.75 acre (0.7 hectare) scenic coastal site at the Scripps Upper Mesa gifted for a peppercorn lease from the University of California, San Diego. It is because of Venter's research track record and his ability to attract funds and people that the Institute has a prominent location in one of the world's richest concentrations of biosciences research. Although the building houses thirty-five researchers, with their own projects and expertise, Venter is critical to its architecture. Just as we have explored the manners by which laboratories construct science, we are interested here in the ways that this laboratory in turn portrays and constructs Venter as a scientist....

J. Craig Venter, the Man

Biotechnologist, biochemist, and geneticist, Venter has twice been named by *Time* magazine as one of the world's one hundred most influential people and has been described as "one of the few great rock stars in life sciences."[1] In 1998, Venter provocatively announced that he was forming a private company that would unravel the genetic code of life. It aimed to do so in just three years, seven years before the projected completion of the state-funded attempt to sequence human DNA, known as the Human Genome Project (HGP). James D. Watson, the HGP's first leader, contemptuously dismissed Venter's sequencing technique as something that "could be run by monkeys."[2] After Venter's company sought to copyright some genetic sequences, Watson claimed that "Craig wanted to own the human genome the way Hitler wanted to

Figure 11.1
Courtyard View of the J. Craig Venter Institute (2013), La Jolla, California. Architects: Zimmer
Gunsul Frasca. Photograph by authors.

own the world."[3] In a more temperate fashion Venter observed, "Watson and I have a lot in common and it's sort of been a love-hate relationship in the early stages."[4] The raw and unabashed ambition Venter displayed then has not abated, nor has opposition to his methods. Having "read" DNA, Venter claims the next stage is to "write" it. Jeremy Rifkin, author of *The Biotech Century* (1998), believes Venter is "trying to short circuit millions of years of evolution, and create his own version of the second Genesis. It's the height of hubris, it's irresponsible, and he can't tell you it's going to be safe."[5] Venter, however, has honed his capacity for risk, as much through his business ventures as through his science. One of Venter's trail-blazing contributions to science has been the design of a business model that twins nonprofit research organizations with for-profit companies. The model aims at the swift translation of scientific discoveries into marketable products for companies, which in turn "donate" funds back to their not-for-profit partners to fuel further research. The J. Craig Venter Science Foundation is his current not-for-profit organization. Launched in April 2002, the foundation merged three of the five not-for-profit research companies Venter had previously led. He personally gave the foundation a USD 100-million-plus endowment that he had amassed from a past venture, Celera.

In 2005, Venter launched a for-profit company called Synthetic Genomics and sought venture capital for its research. He remains its chairman and co-chief scientific officer.[6] Synthetic Genomics (SG) funds 8 percent of the JCVI's roughly 300 researchers and has rights to the intellectual property generated by their research activities. By 2010 the company had an imputed valuation of over USD 500 million, of which Venter owned 15 percent. Around USD 400 million of that came from corporate companies. Synthetic Genomics has three wholly owned subsidiaries: SGI-DNA™, Synthetic Genomics Vaccine, Inc. (SGVI), and Genovia BIO™.[7] SGVI is focused on the development of synthetic vaccines, including a program in partnership with the pharmaceutical conglomerate Novartis—clients of the project discussed in chapter 5. Genovia's mission is to develop "bio-based, consumer friendly products which are sustainable and economically advantaged,"[8] and it received USD 300 million from Exxon Mobil to research and produce algal biofuels (figure 11.2). A fourth subsidiary company focusing on agriculture, Agradis, was co-founded in 2011 with Mexican asset management and investment company Plenus S. A. de C. V. Select assets of Agradis were subsequently purchased by Monsanto in January 2013 with other interests maintained by SG and Plenus in a new company, AgraCast. Coincident with this purchase, Monsanto made an equity investment in SG and signed a multiyear research collaboration. Agradis aimed to develop crop protection and plant growth-enhancing products, particularly for castor and sweet sorghum.[9] If you're not keeping up with this complex network,

Figure 11.2
Craig Venter, standing, celebrating the USD 600 million algae deal with Exxon Mobil at his house in La Jolla, California, July 14, 2009. Photograph by Steve Jurvetson.

then you might need to take advantage of Venter's fourth company, founded in 2014, focused on aging. Human Longevity, Inc. (HLI), was launched despite Venter having once declared that "health care should be the least of our worries," given the graver challenge of climate change.[10] HLI will use genomics and stem cell therapies to find treatments that allow aging adults to stay healthy and functional as long as possible. The company's homepage motto, seemingly without irony, is "Aging is the biggest risk factor for virtually every significant human disease."[11]

The universities, government organizations, investors, and other private corporations that have orbited around the star presence of Venter and his team are a complex constellation critical to the formation and operation of the new JCVI building.[12] Venter's alma mater, the University of California at San Diego (UCSD), initiated the return from Rockville, Maryland, of their most famous graduate in 2005. UCSD offered two sites for the project, one close to the famed Salk Institute for Biological Sciences and the other in the heart of Torrey Pines Mesa with a prospect of the sea. Venter chose the latter. Located on 2.03 acres at the edge of the Skeleton Canyon Ecological Reserve, the new building at La Jolla accommodates 125 scientists and staff. When the building was completed, this constituted just over one-third of JCVI's operational workforce—the others were in Maryland. By 2018, the total workforce was down to around 200, with most staff at La Jolla and the future of the Maryland site in question. The La Jolla building is the only purpose-built accomodation of Venter's multifarious organizations and, in this respect, offered him a unique opportunity to engage architecture as a medium by which to express both the purpose and values of his broader mission. But as we have seen architecture is more complex than this.

Representation

Once accused of "science by press release,"[13] after Venter "tied" in the race to decode the human genome, he was dismissed by Watson as "a good marketer" whose contribution was merely to speed things up.[14] There is some hypocrisy in this accusation, for, as discussed in the previous chapter, Watson has himself courted publicity (and controversy). Venter, too, has published accounts of his life and scientific discoveries including, *Life at the Speed of Light: From the Double Helix to the Dawn of Digital Life* (2014). His early biography, aptly described as a "study in ambition," was titled (again, without irony) *A Life Decoded: My Genome, My Life* (2007).[15] In his review of the book Steven Shapin describes Venter as "aggressive, arrogant and ruthlessly competitive" as well as "belligerent, innovative, ambitious and entrepreneurial."[16] *A Life Decoded* includes details of Venter's genetic inheritance, and consequent health risks, based on what he learned from mapping his own genome. He counters this scientific representation of the self with narrative reflection and recollections of past identities: the navy recruit, mature-aged student, young father, surfer. Questions of agency and "truth" lend frisson to the account.

Venter's third wife, Heather Kowalski, was for a long period his publicist. (She is now JCVI's chief operating officer.)[17] The marketing and communication undertaken through Kowalski's team since 2002 seek to explain and promote his scientific endeavors to a broad audience. However, it is Venter's personality, lifestyle, and ambition that have been most attractive to mainstream news channels. For example, an article on biological patents in *Esquire* magazine (2007) detailed Venter's frenetic work life, while Venter and Kowalski's Californian lifestyle and renovated home in La Jolla were covered by the *Wall Street Journal* (2010).[18] Venter was featured on CBS television's *60-minutes* (2010) and appeared in *The Frost Interview* (2012) for Al Jazeera television. His interview for BBC One's premier episode of *Bang Goes the Theory*, which aired on July 27, 2009, took place on his 104-foot (32-meter) sailing sloop, the *Sorcerer II*. This was the same vessel upon which he and his crew sailed during 2003 and 2004 collecting marine organisms.[19]

Soon after the completion of the JCVI's La Jolla headquarters in 2013, Venter was photographed alone in his office by Michael Lewis for the "Life's Work" interview series in the *Harvard Business Review*. Shortly thereafter he was photographed by celebrity photographer Kurt Iswarienko in a study for the prestige timepiece company Jaeger-LeCoultre. These photographs depart from the image of a man-of-action conveyed by images of Venter in the laboratory and on the boat. They depict him as a contemplative scholar. The photographs of Venter in the laboratory, on the boat, at home, and in the study belong to the genre of environmental portraiture. Environmental portraiture relies on notions of authenticity and naturalizes the relationship between the subject and setting, seeing one as revelatory of the other. It is a genre that conveniently ignores the possibility that a setting is borrowed, contrived, or styled, or that the photograph is manipulated after the shoot. Moreover, as anyone who has hidden evidence of sloth or vice before the arrival of house guests would recognize, the design of the places one inhabits—office or house—is intimately bound to the self one desires the world to see. The design of Venter's office is motivated by its anticipated role in exactly this sort of portrayal—not only through the medium of photography, but also as he engages with employees and visitors. As we will see, the office, along with the JCVI building that accommodates it, works to portray Venter as a very particular kind of scientist and businessman. Understanding these vignettes of Venter is, then, one way to understand the architectural expression of the JCVI and to prise open the more complex representational schema of its architecture and interior design.

The Scientist in His Study

Thinking is invisible. To compensate for this lack of explicative action, the study has long been coopted as an exteriorized expression of internal processes. It can be furnished with artifacts to narrate achievements, personal qualities and life events in a semi-naturalistic manner. For example, take Albrecht Dürer's etching *St. Jerome in His Study* (1514; figure 11.3) or Vittore Carpaccio's painting *St. Augustine in His Study* (1503). Each scene is dense with items—globes and astrolabes, open and closed books, timekeepers and other instruments of measure, a desk, an empty chair, scatter cushions. Both saints are illuminated by daylight from a window to the side and rear of the desk where they sit working. Each has a loyal dog at his feet. St Jerome is also accompanied by a sleeping lion, and, while *Panthera leo* is an unlikely companion (he's part of the personal legend and iconography of this Saint), other items present are the archetypal accoutrement of a life spent thinking and writing. Such items, which are plausible as furnishing, have iconographic significance. These depictions come at a time when

Figure 11.3
Albrecht Dürer, *St. Jerome in His Study* (1514).

man's existence was moved to the center of philosophical concerns, when human agency and individual subjectivity—the fashioning of a life—were both sanctioned and fostered. Personhood was constructed as the outcome of one's own determination and behavior rather than one's destiny, birthright, vocation, or social position. This idea that the self could be invented and performed made portraiture, including self-portraiture, related to the act of self-imaging and performance. In turn, the background of portraiture becomes a stage-set for that performance. Building on the Renaissance concept of the interior as a setting of professional achievement and personal agency, the nineteenth century saw the emergence of the architectural interior as a natural and unavoidable exteriorization of the psychic interior. The idea became powerful in fashion magazines of the twentieth century, which, without irony, guide reader towards the making of lovable rooms so their owners might appear endearing. In the opposite direction, pop psychology instructs readers to see through the artifice of the other in order to decode the gestures, objects, and expressions that escape conscious control—the idea being that the astute viewer of such images can impute personality and state of mind, as well as occupation, social class, and taste, from the room in which a subject is portrayed. Similarly, the idea that interior and inhabitant are mutual self-presentations underpins the two photographs of Venter in question.

In 2015, Venter was featured in Jaeger-LeCoultre's "Open a whole new world" campaign and photographed by Kurt Iswarienko, with "out-takes" by cinematographer Rob Chui (figure 11.4).[20] We are told the campaign images "are a moment caught in time, a story that is just about to unfold as the various protagonists go about their extraordinary lives, unaware they are being captured on film."[21] This, of course, is as preposterous as the presence of a lion in a study, for the production involves a considerable team and is far from candid street photography.[22] The fiction's usefulness, however, is that the effort reveals what the company and their advertising consultants believe is an attractive portrayal of a scientist for their audience. *The Jewelery Editor* describes how, in the photograph, "the American scientist who created the first cell with a synthetic genome, sits in his study, his gaze resting on a distant point, lost in thought."[23] Formally dressed, his face averted from the camera, Venter conspicuously sports the company's Duometre Unique Travel Time watch in rose gold with a brown alligator leather strap. The photograph is captioned "Dr. Craig Venter. Decoder of Human DNA." On an ornate timber desk sits an Atmos clock. Light from a multi-paned, timber-framed window behind Venter softly illuminates the room, imparting a halo while rendering his face in shadow. Dark timber furniture sits heavily on a parquetry floor and the walls are richly paneled, or lined with shelves of books. On a table beside the scientist is a casually folded newspaper and a glass half-full of a transparent liquid (water or gin).

Figure 11.4a
Craig Venter for Jaeger-LeCoultre's "Open a whole new world" campaign (2015). Drawing by author from a photograph by Kurt Iswarienko.

Figure 11.4b
Craig Venter for the Harvard Business Review series Life's Work (2014). Drawing by author from a photograph by Michael Lewis.

A worn, brown leather travel chest occupies part of the foreground upon which rests a single leather-bound book. With its sepia coloring and gentle light, the atmosphere is redolent of a European gentleman's club or the personal library of a scion educated to manage the family fortune.

Venter does not, however, look to be at home. The study in which he is depicted is not his own, but a room that conforms to Jaeger-LeCoultre's vaguely Art Deco retro-aesthetic. In the "behind the scenes" interview, the combative and passionate public persona seen in TED talks and media interviews is subdued and guarded.[24] It is understandable that the luxury brand would reject the laboratory as a site too strongly associated with labor, striving, modernity, and industry. American laboratories have, historically, been places where outsiders—Jewish refugees like Albert Einstein and determined women like Barbara McClintock—succeeded through their intellectual prowess, hard work, and lack of interest in monetary gain. Venter, while wealthy and successful, retains something of his outsider origins—a self-made, brash WestCoast entrepreneur, Vietnam veteran, and surfer. But if the laboratory is unsuitable for Jaeger-LeCoultre's purpose, why not photograph him in his own study at home or work, as McClintock was at Cold Spring Harbor in 1951, or as Einstein was by Ernst Haas that same year?

Indeed, Venter *was* photographed in his actual study—his office at the JCVI—by Michael Lewis for the *Harvard Business Review* series "Life's Work," in 2014 (figure 11.4b). Lewis shows Venter in a relaxed posture on on a contemporary black leather, two-seater couch with his left arm caressing a miniature poodle named Darwin. Two artfully graphic cushions punctuate the black of the couch. He is wearing jeans, a collared long-sleeved chambray shirt with rolled sleeves, and no tie—his regular work attire. He sports a watch but the brand is not visible. Between the couch and the backdrop of windows are a vintage Indian motorcycle and a telescope. In front, on a large round coffee table that obscures his feet and calves, is a bronze sculpture of two entwined hands, several books, two shells, and a dragon-like sculpture. The most arresting object on the table is a 3-D printed model of Venter's brain in a domed vitrine. Despite the presence of the poodle, a lap dog breed favored by aristocratic ladies in the eighteenth century, it is a room that has been styled to convey masculine authority.[25] Here, too, Venter looks away from the camera and into the distance to gaze upon the vista of the Pacific Ocean. This gaze into the distance, beyond the frame, invokes his visionary status. Darwin the poodle, on the other hand, stares vacantly.

While we know from visiting the JCVI that objects have not been introduced into the shoot, art director Annie Chin has edited, selected, and possibly manipulated the image. We do not see Venter's desk to the left of the couch, under which lies Darwin's well-used bed and bowl. On the desk is a 3-D print of just half of Venter's brain, a

Figure 11.5
The executive office at the J. Craig Venter Institute (2013), La Jolla, California. Architects: Zimmer Gunsul Frasca. Photograph by Nick Merrick.

bronze award in the style of Auguste Rodin's *The Thinker*, and a broken model of *The Beagle*, the ship upon which Darwin (the man, not the dog) voyaged (figure 11.5). Charles Darwin's importance for Venter is reinforced by a copy of John van Wyhe's "lavishly illustrated" *Darwin: The Story of the Man and His Theories of Evolution* (2008) on the desk, along with books on science and art, computers and other communication technologies. Shelves of books are interspersed with medals and objects: a crystal bowl, an engraved silver tankard, a ceramic elephant, a class photograph of Venter's Navy cohort, a hat. In one corner is a potted plant and a coat rack holding a single white lab coat, and in the opposite corner a second vintage motorcycle on a plinth, this one a Harley Davidson. A ten-seat boardroom table with chrome and black leather office chairs dominates one side of the room. The walls around the table are crowded with framed awards and certificates commemorating, among other achievements his induction into the American Academy of Arts and Sciences and being made a fellow of the American Association for the Advancement of Science; it also includes an autographed photograph with President Barak Obama on receipt of the 2008 National Medal of Science, America's most prestigious scientific prize. Prominent in its absence is the elusive Nobel Prize.[26]

The photographs for Jaeger-LeCoultre and the *Harvard Business Review* are as laden with narrative signifiers as the fictional inventions of Dürer and Carpaccio. Each tells a different story. Neither story is "true," but not because photography imposes a layer of artifice upon a "real" study. There is no authentic, or unmediated, study for this public man who is conscious of his importance in the history of science. While his office at the JCVI is crammed with displayed objects in the manner of a *wunderkammer*, this plenitude should not be mistaken for intimacy or disclosure. The room is furnished to underline Venter's achievements and assert his passion for risk-taking sports and speed. It is a means of scaffolding and staging a desired persona—the maverick scientist-cum-entrepreneur—and is as accurately targeted at its audience as the Jaeger-LeCoultre "study" was at theirs. Venter's office gains true significance when considered in a wider view—one that takes in the spatial relations between it and the other rooms in the JCVI, as well as other buildings accommodating Venter's organizations, and, from there, other sites of science and business both close and distant.

The Architecture of the J. Craig Venter Institute

Zimmer Gunsul Frasca Architects (ZGF), the architects of the JCVI, were founded in Portland, Oregon, in 1942 and have since expanded to Seattle, Los Angeles, Washington, DC, New York, and Vancouver. The firm is the ninth highest earning architecture firm in the United States in the science and technology sector, with revenue in 2013 of over USD 10 million.[27] ZGF's fees for designing the JCVI in 2013 were USD 726,000.[28] ZGF also declares an interest in sustainability and energy efficiency, ranking at seventeenth in the United States in 2013 for its "Green Building Revenue" by *Building Design and Construction*.[29] Like Centerbrook, the architects for the Cold Spring Harbor Laboratory, ZGF is a recipient of the American Institute of Architects highest honor, the Architecture Firm award. The firm designed so many major buildings in Portland that in the 1990s the city was dubbed "Frascaland."[30] Despite this and the ongoing scale of its work, the practice is little known outside of the United States and has not been the subject of sustained critical discourse.[31] ZGF is a commercial architectural practice that thrives on the consistency of its product and services and its ability to responsibly serve the interests of its clients.

Ted Hyman, a partner of ZGF Architects who led the design of the JCVI, recalls that "Craig Venter came to us with the goal of designing the 'Salk Institute of the 21st Century.'"[32] Venter first came to San Diego in 1971 and "was inspired architecturally and historically" by the then six-year-old Salk Insitute. He recalls, "The impact that Salk had obviously influenced me.... I certainly wanted the architecture [of the JCVI] to be uniquely spectacular in its own way, to make a statement about science and science fitting in with the environment."[33] Both Jonas Salk and Venter are unusual in being able to occupy, while still alive, purpose-built "legacy" scientific premises. At the JCVI the relationship between the client and architect was not, however, like the fruitful partnership that existed between Salk and architect Louis Kahn. Indeed, it might have been expected that Venter would have been more ambitious in his choice of architects and sought out an alliance with a California "starchitect" counterpart, say Frank Gehry, Eric Owen Moss, or Thom Mayne. The process of shortlisting architects and assessing each of their Expressions of Interest (EOI), however, was led by the estates office of the University of California at San Diego, and such offices are notoriously risk averse. Although Venter was involved in the selection and briefing process, the building has none of the monumental character that made the Salk Institute an architectural icon. It departs radically from the Salk's spatial order and consequent social organization.

ZGF's design utilizes the same raw teak walls and exposed concrete as the Salk Institute (the concrete specialist who oversaw work on the East Building Addition in 1995 was employed),[34] but Kahn's design gave the offices of the lead scientists equal size and prominence around a courtyard so as to suggest their democratic engagement as a collective. Venter's office at the JCVI, on the other hand, is singularly large and revealed only from the rear of the building. It sits at the prow of the administration and facilities wing at the farthest distance from the main entry and is the only office with ocean views and a private outdoor terrace. Its distance from the entry dramatizes and prolongs the procession the visitor, or employee, makes for an audience with Venter. It is truly an "inner chamber."

The JCVI's courtyard, unlike that at the Salk Institute, is long and narrow, and its flanking wings are asymmetrical—one curved and three stories high beneath a flat roof, the other straight and a single story beneath a skillion roof. The roof of photovoltaic panels obscures the sky above the courtyard, while Venter's office partly obscures views to the sea. The courtyard's landscaping by David Reed consists of gravel, feature stones, timber paths, and patches of mondo grass laid out in a fussy manner at odds with its attempts to emulate the traditional Japanese garden. The building and its grounds fail to cohere into any single iconic image. No scrums of photographers amass here.

Formal aspirations did not drive the design of the JCVI. Rather its form is the outcome of the primary goal of achieving "a net-zero energy laboratory building."[35] Its design "revolved around energy performance, water conservation, and a multitude of other sustainable efforts."[36] The architects drew on current technologies that reduce dependency on nonrenewable resources. These include orientation, sunshades, high-performance glazing, operable windows, a naturally ventilated car park with bicycle storage, automatic shut-off for unused equipment, variable brightness settings for artificial lighting, chilled beam air-cooling, recycled water for nonpotable water functions, low water landscaping, rainwater collection, and high-efficiency plumbing fixtures. The architects also specified materials with low-embodied energy—high-strength

Figure 11.6
Photovoltaic roof at the J. Craig Venter Institute (2013), La Jolla, California. Architects: Zimmer Gunsul Frasca. Photograph by authors.

concrete with a maximum amount of recycled content, bamboo flooring, and Spanish cedar timber siding. The most visible sustainable feature of the laboratory is the mass of integrated photovoltaic panels—two arrays comprised of 26,124 SF of photovoltaic surface across 1,488 Sunpower E20/327 panels—that the architects predict will exceed the building demand. Given the huge power demands of laboratories, this is an impressive feat. To achieve sufficient area, the array covers the roofs of both wings and the courtyard between them (figure 11.6).

At first glance, and by the architect's own reckoning, the design of the JCVI is largely a technical response to a set of performance parameters against which the building's success can be quantified by flow sensors, thermometer probes, and meters that register energy generation and use. Some of the machinery for achieving Leadership in Energy and Environmental Design (LEED) certification is visible and is clearly intended to signal the JCVI's commitment to the natural environment. For Venter, "The Institute's unique design melds the environmental philosophies of our genomics research with … sustainability goals."[37] In fact, Venter has elsewhere expressed doubts about the mechanical approach to sustainability. He recalls in his memoir, *A Life Decoded*, "I wanted to do more than just using less oil and gas or installing a solar panel."[38] Venter ended up with 1,488 of them at the JCVI and more at home. Without diminishing the fuel savings made here, we insist that the JCVI symbolically needs solar panels because the research has a rhetorical dependency on the recognition of human-caused environmental degradation. The JCVI's research program into synthetic genomics requires the identification of a problem—actually, a set of problems—that synthetic genomics specifically will solve.

Climate change is, as Bruno Latour argues was the case for anthrax in his analysis of Pasteur's practices, "a construction made by statistic-gathering institutions."[39] That is not to say it is "alternative facts"—as the deniers of human-caused climate catastrophe would have it; rather, an active and accurate gathering and analyzing of statistics is required so as to identify global warming, rising sea levels, decreasing resources of fossil fuels, increased production of carbon dioxide, and so on. Without these statistics it is not possible to first diagnose and then to propose, or to make, an intervention. Climate change research makes it possible for Venter to declare that "modern life, in short, is unsustainable" and make the next leap to propose "environmental genomics" as the answer.[40] Self-replicating synthetic genomics and microbes have many potential uses—only one of which is the engineering or bioremediation of the earth's "sick atmosphere."[41] Through bioengineering, Venter's broader crusade is to "change the way we produce everything from food, to medicines, to fuels and clean water—even building materials for housing."[42] In short, the design of the JCVI participates in pitching climate change as a problem to be solved by advances in science and technology, rather than plant-based diets, silvopasture, educating women, family planning, refrigerant management, walkable cities, or any of the solutions proposed by Drawdown that, instead, require political will, behavior and policy change, social or economic revolution.[43] It situates Venter and his team as the scientists who would save the world through genomics research into alternative fuels, crop improvements, and species modification. The JCVI, while it expresses the sustainable goals and values of its namesake in familiar ways, euphemizes the contentious nature of the work done within its walls. In the same manner, the spatial independence of the not-for-profit JCVI belies the intellectual property arrangements, and entanglement in, Venter's for-profit ventures.

Venter's BioSciences Empire

While the JCVI's architecture declares the organization's commitment to environmental sustainability by minimizing its draw on nonrenewable resources, Venter, it is reported, enjoys "doing large donuts" in his 45-foot speed boat.[44] In addition, he owns a fleet of vintage and new vehicles, including a 1939 Bugatti motorbike, a 1905 Reo motor car, a Range Rover, and a 1996 Aston Martin.[45] By way of concession, he also owns an electric powered Tesla, which he uses to "balance out the others."[46] Likewise, and perhaps to balance out the JCVI's sustainability performance, Venter's for-profit partner organizations reside in leased accommodation, the architecture of which is oblivious to even the simplest fuel-reduction strategy. Synthetic Genomics, for example, leases from Nexus Properties part of the Nexus Center in the Torrey Pines Science Park, three miles from the laboratory of JCVI.[47] Built in 1989, the two-story speculative development in which the sixty researchers employed by Synthetic Genomics are housed is fully air-conditioned with unscreened reflective glass curtain window walls. The *New York Times* describes it as a "humdrum ... box plopped beside a highway."[48] At Rockville, Maryland, the JCVI occupies buildings of similar ilk constructed in 2004 and leased from BioMed Realty Trust, which also owns the property that Human Longevity, Inc. (HLI), occupies at 10835 Road to the Cure, San Diego (figure 11.7). (The irony of a road for gas-thirsty, carcinogenic vehicles being named this way is not lost on us.) Built in 1978 and renovated in 2007, HLI occupies a two-story box surrounded by an asphalt apron of car parking. The office headquarters at 4570 Executive Drive, La Jolla, are

Figure 11.7
Human Longevity, Inc., 10835 Road to the Cure, San Diego. Photograph by authors.

housed in a speculative development built in 1999, with an ostentatious hemispherical entry portico, reflecting pond, and obelisk. (Each of these sites engages private security guards who intervened in our attempts to take photographs from the street.)

Such contradictions, and the "twinge of guilt" *Wall Street* journalist Sara Lin detects when Venter shows her the toys in his garage, are familiar tropes of the First World.[49] Rather than dismiss the JCVI's sustainable architecture as greenwash, the more interesting question is how the representation of environmental commitment works in tandem with the building's emphatic staging of Venter as its chairman and chief scientist. The question then is not who is Venter, nor what is his architecture, as if there were an authentic truth to be revealed. It is what the images and the architecture make possible that matters. There are three key audiences for Venter and his architecture. First, there are his scientific collaborators who appear in different roles across the for-profit and not-for-profit companies. Microbiologist and Nobel Laureate Hamilton Smith, for example, is a JCVI employee, and also a co-founder, board member, and shareholder of Synthetic Genomics, Inc. Smith had been on the scientific team at the Institute for Genomic Research, Venter's nonprofit firm involved in decoding the human genome. JCVI employees Ken Nealson and Clyde Hutchison are also members of the scientific advisory board of Synthetic Genomics. JCVI board members Alfonso Romo, Theodore Danforth, William Harman, and Dean Cornish are all shareholders of Synthetic Genomics. JCVI's president, Karen Nelson, is simultaneously head of Microbiome for HLI. Kowalksi Communications, the company owned by Venter's wife, is the media contact for Synthetic Genomics, JCVI, and HLI. One could go on. The point is that Venter has an intimate and loyal group of trusted collaborators simultaneously managing the business of both the for-profit and nonprofit organizations. Most worked with him long before the new building and are not in need of being persuaded of the value of his research or his legitimacy as the organization's visionary leader.[50]

The second group are high-achieving entrepreneurs who, like Venter, possess scientific knowledge and business acumen, such as his two co-founders in HLI, Dr. Robert Hariri and Dr. Peter Diamandis. Hariri is also a member of the board of trustees of the JCVI. Hariri is a celebrated surgeon and biomedical scientist whose company, Lifebank USA, a placental and cord blood banking business, was acquired by HLI in January, 2016.[51] Diamandis has an undergraduate degree in Molecular Genetics and a graduate degree in Aerospace Engineering, both from MIT, as well as an MD from Harvard Medical School. Founder of the X Prize Foundation, known for its USD 10 million Ansari X Prize for private spaceflight, Diamandis is also co-founder of the business incubator and conference hub Singularity University, and a company designing spacecraft to enable detection and mining of asteroids for precious materials called Planetary Resources.

Juan Enriquez, one of the three co-founders of Synthetic Genomics, has an MBA from Harvard and no scientific expertise, but was the founding director of the Harvard Business School's Life Sciences Project. He is the author of numerous books extolling the virtue of a post-natural world predicated on a synthetic biology economy.[52] Convinced by his own predictions, Enriquez founded Biotechonomy Ventures and invested in BioTrove, Xcellerex, and Synthetic Genomics, before heading to a role as managing director of Excel Venture Management. People like Diamandis, Hariri, and Enriquez recognize that high risk can yield high return. The adrenalin-fueled risks conjured by Venter's "outlier" approach to science, his Darwin-inspired quest on the *Sorcerer II*, even the vintage motorcycles in his office speak to the predilections of this group who see a kindred spirit in Venter's "winner takes all" approach to life, science, and entrepreneurship. The Navy cohort photograph in Venter's office serves to qualify the virtue of these traits, offering a subtle reminder of his uncertain beginning, resilience, and determination.

Which brings us to the third group, the corporate investors. The Malaysian conglomerate Genting and its chief executive, K. T. Lim, together own nearly 20 percent of Synthetic Genomics, making them the largest holders.[53] Lim has also privately contributed USD 70 million to HLI. K.T. Lim's interests are even broader than those of Diamandis, Hariri, and Enriquez. Genting Group is involved in leisure and hospitality, power generation, palm oil plantations, property development, biotechnology, and oil and gas. Genting has partnerships with Universal Studios, Hard Rock Hotel, Premium Outlets, and the Mashantucket Pequots, with whom he established the Foxwoods Resorts Casino in Connecticut, the largest casino in the United States. The scale of investment and ambition of this third group, to which Lim belongs, along with corporations such as Exxon, Monsanto, and Novartis, are much higher and implicate large numbers of shareholders. Their interests must be balanced against expectation of return. These investors may recall the financial disaster that saw Venter voted out of Celera in 2002, the very company in which he had been installed president four years earlier. Investors poured billions into Celera as it climbed from USD 7.20 a share in mid-1999 to USD 247 nine months later. Less than a month after its peak, Celera's stock fell to USD 92 a share and by the end of 2000 it was at USD 36.[54] The shift occurred on March 14, 2000, when the prime minister of the United Kingdom, Tony Blair, and the president of the United States, Bill Clinton, made a joint statement calling for information about the human genome, or genetic code, to be made publicly available. Biotechnology stocks declined more than 12 percent in one day. It remains a volatile and immature industry.

In his detailed account of Synthetic Genomic's commercialization efforts, *Synthetic Biology: Science, Business, and Policy* (2011), Lewis Solomon views the company's collaborations as serving Synthetic Genomics "to validate its technology,"[55] yet there is obviously some ideological room between using "living systems to increase our chances of survival as a species" as Venter puts it, and meeting the purely financial demands of company shareholders.[56] Venter claims that "whoever produces abundant biofuels could end up making more than just big bucks—they will make history.... The companies, the countries, that succeed in this will be the economic winners of the next age to the same extent that the oil-rich nations are today."[57] There are anticipated economic wins, too, for Monsanto and Novartis in their investments in SGVI and Agradis. The CEOs and agents of these companies require a different kind of reassurance than those whose knowledge of the JCVI's science convinces them of the value of supporting the organization or any of the for-profit ventures. Synthetic Genomic's corporate investors need to be confident that its research operations are stable, yet breaking new ground. They need to be sure of Venter's vitality and grasp on its operations and of his ongoing standing in the scientific community. They need visual and material assurance of Venter's aphorism, "If the science works, the business works, and vice versa."[58]

The histories of Venter's corporate partners are not stainless in relationship to human rights and health, environmental sustainability, financial transparency, and business ethics.[59] It is not an easy game negotiating the territory between saving the world through science and doing business with others whose motivations differ. Venter needs support for research that may, indeed, mitigate anthropogenic climate change, but sources with sufficient funds to aid and abet his efforts do so with self-interest and chequered track records. The JCVI building, and Venter's office within it, are symptomatic of these conflicts in that they combine the pragmatism of a technically oriented "green" building with the symbolic aggrandizement of its chairman. Venter's office both registers and constructs these tensions, as it generates niches and surfaces that can be laden with "evidence" of its occupant's legitimacy at the seat of a biosciences empire.

This "evidence" is no more real than is the fictional Jaeger-LeCoultre study; rather, it consists of a myriad of unstable representations that themselves are mediated by photographs and architecture. The photographs in the *Harvard Business Review* and Jaeger-LeCoultre campaign construct and advance a persona at a careful distance from the scientist-businessman whose discoveries threaten the status quo. A slippery semantic operation is required for him to be re-presented as a benign, but accomplished, thinker and gentleman aligned with Jaeger-LeCoultre's world of heritage and nostalgia. He must be recast as "Decoder of Human DNA" in the past-tense, rather than as the man who, in the present, aggressively pursues the remaking of life and the replacement of oil as the world's fuel source. He must be presented as a man of knowledge, not action and consequence. Venter becomes useful for Jaeger-LeCoultre only insofar as he is not himself, or at least not the maverick-self of his youth.

For the readers of *Harvard Business Review* he is similarly humanized, softened, taken out of the laboratory and placed onto the couch. Each of these men is a cipher that is and isn't Craig Venter. His "real-life" office and the JCVI that houses it function as avatars of the "great man of science," who, long after his brain ceases to operate, will continue to exercise his work and exert his influence in a disembodied and spatially distributed way. Their true value lies in the fact that they already do this, regardless of whether the figure behind the myth is really there or otherwise. The research activities at the JCVI and of Venter's previous organizations are intensely collaborative. While Venter is the lubricant for this collaboration and the lynchpin for its business operations, the architecture overstates his centrality because it is easier for the public and for investors to place their hopes upon a singular genius than a faceless organization. While scientific practice is always collectively achieved and discovery is the hard-won aggregation of multiple people and their many insights, the narrative of individual genius prevails. In this sense, the architecture, the interior, and its contents play to this expectation.

12 Investments

Unlike the laboratories discussed in previous chapters, our case studies here—the Alexandria Center for Life Science (2011, ongoing) and the Harlem BioSpace (2014)—were designed with God's instruction to Noah (or to Kevin Costner's character in *Field of Dreams*) in mind: "Build it and they will come." Both are located in Manhattan and are ingredients of a much larger project initiated and subsidized by New York City to realize a biosciences "ecosystem." They are also part of the rapidly growing laboratory real estate industry focused on the acquisition, design, construction, and management of laboratory facilities for lease to prospective tenants. In the last decade the corporations that develop and manage commercial laboratory buildings, such as BioMed Realty, Alexandra Real Estate Equities, and Nexus Properties, have substituted their homogenous product for nuanced developments with architectural pretensions and lifestyle amenities. Where the Alexandria Center is a slickly styled waterfront campus intended primarily for large and established pharmaceutical and biosciences tenants, Harlem Biospace is a co-working space aimed at attracting entrepreneurial start-ups.

The architectural expression and the intended occupants of this chapter's case studies are so clearly aligned that they are as obvious as a hive might be to a bee or a lodge to a beaver. And as with hives and lodges, bees and beavers, the question of which comes first is moot. Speculative laboratories are designed with a type of scientist tenant in mind, at the same time as scientists are then shaped to that ideal. The practices of investment—financial and managerial—in human resources being made across the life sciences are on a continuum with those made in physical assets such as property and buildings. As a set of levers and mutually constructing resources, these investments operate beyond local stakeholders and profits to create "worlds" and modes of life. The rise of speculative laboratory real estate encourages us to explore the nexus between property speculation, intellectual property production, and the production of the twenty-first century scientist.

Shaping Scientist Identities

In previous chapters we have repeatedly heard the claim that scientists are finally being brought out of the miserable confines of their solitary basement laboratories. We have seen, too, how the new laboratories are touted as spaces of sociability, connection, communication, and community. Indeed, Nigel Thrift observed that the encouragement of creative sociability had already become an "established convention" in laboratory building in the 1980s.[1] Thrift credits these ideas to the architectural rhetoric that had grown around workplace design. A repertoire of strategies had emerged that mirrored concepts developed in innovation management theory—hot-desking, smart working, brainstorming, trading ideas, connectivity, flat organization. Architects such as Francis Duffy, whose practice DEGW specializes in workplace design, were quick to exploit the changes taking place in management theory and to argue that design had become essential for corporations using capital investment as a lever to affect organizational change.[2]

Architecture's success in influencing social behavior in the workplace comes, however, not because it actually has the coercive power to mold passive (or unwilling) subjects. Rather, the success of architecture is because the creativity and socializing today's scientific workplaces presume to facilitate is in resonance with the ambitions and ideals of scientists themselves. Again, which came first is beside the point. The contemporary laboratory emerges alongside the "subjective processes" that Maurizio Lazzarato argues are a key aspect of the mutation of all labor as "immaterial labor."[3] Toward the 1980s, a new mode of control began to appear, which was less intrusive than Fordism or Taylorism since it gave workers a certain freedom to manage their time and organize their own activities. By the late 1990s, Taylorism, with its focus on imposing a fixed pattern of movement from above, had been supplanted by team-based structures aimed at "constant improvisation in work organization and the *unobtrusive orchestration of employee values*."[4] In place of compliant bodies, the new high-involvement workplace sought supple minds—commitment supplanted control. However, commitment and creativity remain a "problem" for management because researchers and scientists see themselves as playing a critical role in society. They "demand different kinds of authority relationships" and "pose unique challenges to leadership."[5] The creativity of the worker is a quality integral to the success of the enterprise, but requires a different culture of management. The change is particularly pronounced in the life sciences, where the backdrop of increasing cross-traffic between universities and the private sector has seen pressure on industrial organizations to provide the managerial practices that the academy has traditionally valued. These include autonomy, collegiality, control over research, individual recognition, opportunities for continuous learning, and the support of scholarly publishing so scientists can maintain prestige and prominence in the field.

As Lisa Adkins and Celia Lury argue, contemporary organizational management seeks a "self-transforming subject"—that is, one "who can constantly adapt his or her performance of self-identity and who can claim the effects of that performance of self-identity to define their own goals ('to fill in the blanks along the way')."[6] The self-transforming subject is "the ideal subject of the employment contract."[7] One would imagine that in science the "blank" to be filled is knowledge and discovery, but according to Lazzarato, in the knowledge economy "it is never an individual who thinks."[8] Value is no longer assigned to an author as an agent of wealth creation, although this idea is deeply embedded in scientific rhetoric. As Mackenzie Wark notes, capitalism does not need that much knowledge to function; what it does need are desiring subjects. Hence, subjectivity is a key commodity to be produced.[9] Lazzarato summarizes the situation:

> To work within a contemporary organization means to belong, to adhere to its world, to its desires and beliefs.... It represents a radical change in the "subjectivity" of the organization and the subjectivity of the workers.... on the one hand, it affirms workers' autonomy, independence and singularity (individual substance), on the other hand, it requires workers to belong to the organizational world, since this world is internal to the situation and conduct of the subject.[10]

The formation of subjects cannot be considered autonomous from the assemblages of markets, machines, and matter that produce trans-individual effects. Whereas for Karl Marx the production process made commodities that were to be consumed by subjects, Lazzarato and Félix Guattari diagnose the contemporary situation as one in which the production process makes subjects and regulates desires. Capitalism produces individual subjects within pre-formed identities—boss, scientist, entrepreneur, woman, youth—at the same time as it de-subjectifies and fragments us into mobile component parts of a bigger assemblage, for example, as data. Guattari explains, "It is not the facts of language use nor even of communication that generate subjectivity. On some level, subjectivity is manufactured collectively just like energy, electricity or aluminium."[11] Lazzarato elaborates on Guattari's thesis:

> The production of subjectivity involves expression machines that can just as easily be extra-human and extra-personal (systems that are machinic, economic, social, technological, and so forth) as they can be infra-human and infra-personal (systems of perception, memorization and idea production, sensibility, affect, etcetera).[12]

The laboratory is just such an expression machine. It brings together spatial settings and architectural forms with contractual and funding arrangements, management performance tools and processes, and recreational activities, amenities, and other aspects of soft culture. We do not assume that the developers of biosciences real estate industry read Lazzarato or Guattari, yet their understanding of the ways in which the subjectivity of their clientele can be modulated, and desire for their product created, is both instructive and disturbing.

Alexandria Centre for Life Science

The Alexandria Center for Life Science is in Manhattan's Kip Bay neighborhood (just above the East Village). Opened in 2011, the complex consists of two sixteen-story towers referred to as the West and East Towers, offering a total of 720,000 square feet (67,000 square meters) (figure 12.1). A third tower is proposed for construction. Predictably, the overall "openness of the campus's design fosters the collaborative environment sought by the life science community."[13] No mention is made of the architects, Hillier/RMJM, in any of Alexandria's marketing material.[14] Instead, its selling point is the East River views and the epicurean delights that are to be savored in two on-site restaurants, Riverpark and Riverpark Farm, "curated" by the celebrity chef Tom Colicchio. It also has a fitness center and an acre of open green park with a river front esplanade, as well as a large conference and event center. The glass-fronted, double-height lobbies in two of its three planned towers have dark stone floors and are furnished with dramatic "Spill" benches (27 feet long) in steam-bent white oak (figure 12.2). Designed and produced by Brooklyn-based Matthias Pliessing, the same curvaceous benches can be found in the foyer of the Harvard Business School and the Smithsonian American Art Museum. Echoing the forms of these benches are spiral staircases, of course, reminding us of the structure of DNA.

Alexandria Real Estate Equities claim to develop "collaborative environments" for the science and technology sector that enhance the recruitment of "world-class talent" and inspire "productivity, efficiency, creativity and success."[15] They boast of having pioneered this niche in 1994, with a string of laboratory campuses in innovation cluster locations such as Boston, San Francisco, San Diego, Seattle, Baltimore and Research Triangle Park (Raleigh, Durham and Chapel Hill). The move to New York was made in the context of the city's ambition to take a share of the profits to be gained from the translation of biosciences research to market.[16] The city's former mayor, Michael Bloomberg, set out a range of initiatives designed to reverse the exodus of researchers in the biosciences to Boston and Cambridge. Bloomberg announced in August 2005 the selection of Alexandria Real Estate Equities to develop the East River Science Park into the largest commercial biosciences center in New York City. The 3.7 acres

Figure 12.1
The two completed towers of the Alexandria Center for Life Science (2014), New York City, are architecturally indistinguishable from other office buildings for lease. Architects: Hillier/RMJM. Photograph by Susan Bednarczyk.

Figure 12.2
The foyer of the Alexandria Center for Life Science (2014), New York City, with "Spill" benches.
Photograph by Susan Bednarczyk.

(1.5 hectares) of land on which the development is located was part of the Bellevue Hospital Center and is city-owned and leased to Alexandria. The city provided about USD 13.4 million in capital funds for its construction and the State of New York gave USD 27 million for infrastructure work in connection to the project. The city's leading business group announced a commitment of USD 10 million toward the USD 700 million project through its New York City Investment Fund, which by 2009 had reached USD 15 million. At the launch Bloomberg prematurely boasted, "Building this world-class facility will help us attract top companies, create valuable jobs and make our City the bioscience capital of the world."[17] Also helping to attract companies was the designation of the "Empire Zone," which allowed tenants to qualify for a variety of state tax incentives based on investment, job retention, and employment growth.[18] These benefits were enhanced by the passing of a bill in the state legislature in 2009 for the city to create a USD 3 million biotech tax credit for qualified emerging technology companies.

Alexandria Center's rents, though, are out of reach for most biosciences start-ups and spin-offs and in its first few years attracted only established companies. By 2015, its tenants included Pfizer, Inc., Eli Lilly and Company, Roche Holding AG, Cellectis, Intra Cellular Therapies, Kallyope, Kyras Therapeutics, Lycera Corp, and TARA Biosystems. While this might seem like a success, New York has sound business reasons for being unsatisfied with simply domiciling large pharmaceutical companies while new ideas and young biomedicine researchers seek opportunity elsewhere. The costs of undertaking biomedical R&D has been rising dramatically over the past fifteen years, with the prospect of returns decreasing. Eli Lilly believes that a key strategy to reducing the costs associated with developing new drugs is to shift R&D away from the wholly owned and operated enterprise to one "that is highly networked, partnered and leveraged."[19] Such network configurations would accord scientists greater autonomy, but they also transfer risk to smaller companies and individuals. In other words, strategic alliances, joint ventures, seed investments, partnerships, and outsourcing would all see scientists have greater control over their labor, but less income and career security. For the larger companies to survive and prosper the New York biosciences "ecosystem" requires places for smaller companies to spin off from universities and, ultimately, to feed (or become one of) the larger corporations. The ecosystem needs talent, and such talent requires its own laboratory space.

Harlem Biospace and Epibone's Nina Tandon

With the above in mind, a smaller venture was also initiated by the Bloomberg administration and quickly brought into being as a "correction" to the Alexandria Center.[20] New York would champion an entrepreneurial culture among its academic research and

Figure 12.3
Harlem Biospace (2014) in the Sweets Building, New York City. Photograph by Robert Hackett.

clinical care communities, including the creation of new incubator space and training programs for biotech entrepreneurs. Harlem Biospace is one such initiative. Granted USD 626,000 from the New York City Economic Development Corporation (NYCEDC) in 2013, Harlem Biospace opened a year later. In the old Factory District of Harlem, on the ground floor of the Sweets Building, the handsome five-story red brick building constructed in 1936 was fittingly once a confectionary research laboratory (figure 12.3). It sits between a taxi depot and the offices of the African Services Committee. Further sponsorship was gained from the legal firm WilmerHale, the imaging equipment company Olympus, the German biosciences product giant Merck Millipore, and Amazon. A 2,300-square-foot (213-square-meter) co-working space for biotech start-ups, the site has twenty-four workstations and housed an initial batch of seventeen start-ups. In 2018 rent was USD 995 per month for a desk and a shared wet-lab bench with a six-month lease, Wi-Fi and laboratory equipment included. Harlem Biospace promotes the affordability of its shared wet-lab space and its proximity to Columbia University, the City College of New York, and "multiple Starbucks," as well as its accessibility by train, bus and the Hudson River bike trail.[21]

Of its architecture and atmosphere, its executive director explains, "We've tried to make it different than another lab."[22] That is vague, but let's see how the intention translates. The renovations were designed by Manhattan-based BAM Architecture Studio (BioMed Realty Trust is one of its major clients), although its input focused on the wet laboratory spaces. The space is a ground-floor glass-fronted shop front, with a fit-out that preserves the building's original ceiling height and industrial character. There are raw timber floorboards and exposed services and structure. Furnishings were procured from local industrial designers and furniture makers, and these individuals are featured on the Harlem Biospace page, while the architects are not. There are benches upholstered in vintage Army blankets; desks, chairs, and shelves made of reclaimed wood and steel; the ceiling is adorned with a chandelier of Edison light bulbs; and potted plants sit on the window sills (figure 12.4a and 12.4b). On the wall in the foyer-cum-lecture-cum-social-space hangs an annual clock (one that takes a year to revolve) by local artist Scott Thrift. Harlem Biospace has an active Twitter feed, an educational outreach program, and its own design and science store where one can buy a bust of Nikola Tesla and a T-shirt emblazoned with "Rock Star of Science." Harlem Biospace reports having been overwhelmed by applications to its spaces, which in its first round were tenanted to companies including Biogelx, Exsponge, Cytodel, GrayBox, IrOs, and EpiBone.[23]

If laboratories "construct" desire, then it is worth looking at who, exactly, the scientists are whom the hipster chic of Harlem BioSpace attracts. Representative of the occupants is Nina Tandon, who with Sarindr Bhumiratana, co-founded EpiBone (figure 12.5). Using the adult stem cells of patients, the company aims to grow bone cells from outside of the body for personalized living bone grafts. Tandon and Bhumiratana met as postdoctoral fellows sharing a laboratory at nearby Columbia University.[24] Tandon subsequently studied for her MBA so that she could, in her own words, *transform* "from a biotechnologist to a biotech leader."[25] The transformation has been effected through the active recruitment of publicity, a shrewd understanding of the need to tell a compelling story about science, and the clever deployment of her status as a photogenic young woman in a sector dominated by less photogenic aging men. The company, its mission, and Tandon have been reviewed in the *New York Times*, the *Huffington Post*, *Scientific American*, TedMed, CNN, and the BBC, as well as a host of other media venues. Tandon was a recipient of Fast Company's 100 Most Creative People in Business (2012), CCN's Tech Superhero (2015), and Crain's 40 Under 40 (2015), chosen as one of the World Economic Forum's 2015 Technology Pioneers, and is a three-time TED (Technology, Entertainment, Design) speaker. Tandon is one of a stable of digital innovators of her generation represented by the Lavin Agency, which also

Figure 12.4a
Shared workspace at Harlem Biospace, New York City, with benches by Coil and Drift and lamps
by Allied Maker. Courtesy Harlem Biospace.

Figure 12.4b
Reception area at Harlem Biospace, New York City. Photograph by Vince Ponzo, courtesy of Harlem Biospace.

represents Neri Oxman (architect, designer and once rumored girlfriend of actor Brad Pitt, himself the designer of a coffee table with a 24-karat gold-plated helical base). In January 2013, Tandon was photographed by Evan Kafka for *Wired* magazine holding a model of a heart to her chest in a pose reminiscent of Our Lady of Sorrows. The contrast between this image and the depictions of fellow scientist Craig Venter discussed in the previous chapter could not be more striking.

Figure 12.5
Nina Tandon, Sarindr Bhumiratana, and Mike Lamprecht from Epibone in the lab at Harlem Bio-Space. Photograph by Carla Tramullas.

For TEDMED, Tandon promotes life as "an entrepreneurial journey."[26] Questioned about her future in science, her response was she was "worried about the job I'm going to create! If you think like an entrepreneur, you're never going to be out of work."[27] The entrepreneurial (MBA) version of subjectivity and autonomy is in stark contrast to that of Lazzarato and Guattari. And although this narrative is intensely felt, it is contextualized by an increasingly fiscally restrained academic sector that previously would have offered Tandon and her peers a secure income and career. We note that as an adjunct professor of Electrical Engineering at Cooper Union, she is among an increasing proportion of people whose relationship with the academy is peripheral and tenuous. At the same time, the growing commercialization of research activities and valuing of entrepreneurial behavior within universities make her decision seem less of a departure from the academic route. Tandon's optimism belies the significant risks she has taken.

EpiBone launched with USD 350,000 from Breakout Labs. Breakout Labs was a program set up by PayPal co-founder Peter Thiel's personal foundation to "jailbreak" academic research in science into the market economy. While Thiel's grant is not to be sneezed at, it clearly is not enough to sustain the project. While still at Columbia University, the team had used USD 10 million in federal grant money for its early research without even reaching animal trials. Assuming that everything goes well with those and subsequent human trials, EpiBone's technology will reach the market only in 2022 or 2023.[28] Very significant cash will be required to keep its team and rent paid and the work progressing for almost a decade. When it was incorporated in 2013, Tandon cautioned her investors not to be hasty or expect a return. "We're slow and steady, we're science nerds, and we are aiming to help humanity."[29] EpiBone raised an impressive USD 4.2 million from sixty-six investors, but will need a second series of investment.

When James Watson hosted Gala Dinners to convince New York City's establishment to fund cancer research, he allegedly untied his shoelaces to convey the impression of a man preoccupied with higher thoughts. But it's a very different image of science nerdiness that the team at EpiBone presents—cosmopolitan, urbane, flexible, multicultural, and youthful. EpiBone's Internet homepage details their impressive qualifications, as well as their personal bodily investments. Little of what is presented is surprising; indeed, it is a wholly consistent image. Tandon describes herself as enjoying "yoga, rock climbing, skiing, surfing, marathon running, & playing with her nieces."[30] For Bhumiratana, its "rock climbing, snowboarding, & soccer." Tandon is acutely design conscious and well networked in the art, design, and architecture scene.[31] Interest in the synergy between biology and design prompted her to collaborate with architect Mitchell Joachim on the book *Super Cells: Building with Biology* (2014) and to appoint artist-in-residence Maia Yoshida in 2015. Yoshida has been making jewelry alongside the scientists out of decellularized bovine bone.

Situating Tandon and EpiBone in New York's cultural milieu involves an entirely different set of social media pages from those we trawled while researching the philanthropic and social context for the Cold Spring Harbor Laboratory. In August 2016, Tandon married Noah Keating in the Louis Kahn–designed FDR Four Freedoms Park on the southern the tip of Roosevelt Island, home of another Bloomberg project to stimulate scientific innovation, the Cornell Tech campus. Keating is a technologist whose company, Mathbeat Industries, blends mobile device experiences, live events, and interactive installations. The pair defied convention by donning top-to-toe black for the wedding ceremony, while for the reception, Tandon, whose ethnicity is Indian, donned a sari and Keating a kurta. Their wedding registry gifts included an "Organic Olive Wood Cutting Board with Leather Handle," a donation to Four Freedoms Park, and a botanically inspired painting from Gerda Van Leeuwen.[32] Their duplex on Roosevelt Island is floored with recycled timber from the Domino Sugar Factory in Tribeca.[33] Tandon's Pinterest board is almost entirely made up of photographs of interiors, furniture, design products, and architecture.[34] From this Google-stalking we understand that scientists are not immune to the cultivation, curation, and publication of their

"lives" via social media. More to the point Tandon's socializing, like that of her contemporaries, far exceeds the gregarious scientist envisaged as a person who has coffee
regularly with a colleague in the laboratory across the way. Tandon's networks extend
well beyond the walls of the laboratory to include other creative professionals such as
architects, artists, technologists, journalists, academics, and angel investors.

In the face of architects who view the laboratory as a technology for inciting employees to socialize, Tandon's self-fashioning and her apparent freedom from imposed
managerial strictures present something of a conundrum. Her narrative is one of self-
determination. Yet, the subject position of the hip biotech entrepreneur preexists Tandon and others—it is produced and reified through government policies, university
degree programs, mentoring, popular media and culture, as much as through architectural settings. Initiatives such as NYCEDC's Early Stage Life Sciences Funding Initiative
and Thiel's Breakout Labs invite individuals to shape themselves as candidates.[35] The
MBA Tandon earned and new programs such as the Technion-Cornell Institute "Runway Startup Postdoctoral Program" are institutional settings in which individuals are
ushered "from an academic mindset to an entrepreneurial outlook."[36] The hipster chic
laboratory environment of Harlem Biospace—from its uptown location to its recycled
furniture—materializes this outlook.

Since 2010, large pharmaceutical companies have also been looking to strategies
that might harvest youthful energy. Companies have been opening "incubator" spaces
for lease to biosciences start-ups—Johnson & Johnson have opened five "JLABS" across
the United States housing over one hundred science start-ups. Bayer Corporation and
Merck have followed. So, too, have the biosciences realtors. Which brings us back to
the Alexandria Center. Alexandria opened one space under The Science Hotel® brand
and another under its Accelerator Corporation subsidiary at its New York campus. The
common room in The Science Hotel® at Alexandria Center for Life Science makes a
stark contrast with the sober boardrooms on other floors, wherein black leather and
chrome Eames "Time-Life" chairs sit around large mahogany tables.[37] The Science
Hotel® is furnished with a bright rag rug, strings of bare bulb lights, and wire chairs in
primary colors around raw timber trestle tables. Publicity photographs show a single
game of "Connect Four" as the table centerpiece (figure 12.6). First sold in 1974 (and
falsely rumored to have been invented by David Bowie), it's an abstract strategy game
that the target market for biosciences incubator spaces might recall from childhood.

In mid-2016 Alexandria announced that 15,000 square feet (1,400 square meters)
of its New Manhattan campus, roughly half of the fourteenth floor of the West
Tower, would open the following year as incubator space under a third brand, Alexandria LaunchLabs™. Marketed as affordable, full-service "plug-and-play" life science

Figure 12.6
The Science Hotel meeting room at Alexandria Center, New York City. Photograph by Hollis Johnson for *Business Insider*, courtesy of Business Insider/Wrights Media.

laboratory space, it offered tenancy through competitive application. The Internet
homepage of Alexandria LaunchLabs™ is the graphic equivalent to the interiors of
The Science Hotel ®, with bright graphics on a black background and images of young
scientists at work and play. Alexandria's slick foyer and white-paneled elevators—so
prominent in its marketing to large corporations—are nowhere to be seen. Instead
a different set of attractions are featured. Alexandria is referred to as a "collaborative
urban campus" with "cutting edge amenities," including "on-site wellness coaches"
and "mentorship."[38] Young "scientists" are shown enjoying yoga in the gym.

"From single-bench scientists to multinational pharmaceutical companies, Alexandria is committed to supporting life science companies at any stage in their discovery
and translation of groundbreaking medical innovations," says Catherine Nuccio of
Alexandria Real Estate Equities.[39] With office spaces starting at USD 1,995 (more than
twice the rent at Harlem Biospace for a workstation and lab bench space), this might
be a bit of a stretch, but by 2018 Alexandria claimed that LaunchLabs® was now home
to twenty life science start-ups. As an incentive, Alexandria annually award the USD
100,000 Alexandria LaunchLabs® Seed Capital Prize. In 2018, the prize went to a Cold
Spring Harbor Laboratory spinoff company called MapNeuro. The ecosystem of cities,

companies, start-ups, spin-offs, and scientists is a tangled one not confined to any bank, least of all that of the East River.

Biosciences Real Estate

The biosciences real estate industry is driven to actively elicit desire for its products, but to succeed, it needs to do so in a way that allows scientists to retain a sense of self-determination. As Deleuze and Guattari remind us, desire is "never an undifferentiated instinctual energy, but itself results from a highly developed, engineered setup rich in interactions."[40] A sense of autonomy, ambition, and flexibility places the scientist in harmony with the image the biosciences project (and, not coincidentally, with the modulation of subjectivities necessitated in capitalism). But it is easy for those who seek to exploit the desires of the scientist-cum-entrepreneur to get it wrong. The cynicism around Alexandria's offering to start-ups, with its infantilizing interior décor and graphics, is palpable. In view of criticism of its exclusivity and unaffordability, Alexandria's CEO Joel S. Marcus defended his company's pricing and policies on the grounds that "we want the companies to be world-class. We don't want anyone diddling around."[41] If this statement seems to contradict the obvious diddling around connoted by yoga and Connect Four, Alexandria's marketing manager did not come to the defense of her boss. Jenna Foger proposed that "it is our collective responsibility to solve the funding bottleneck that local seed-stage companies currently face in order to catalyze innovation and strengthen this important segment of the New York City community."[42] Satisfying this *collective responsibility* (to which Alexandria was already contractually committed in its bid to develop the land) will take up just 2 percent of the lettable floor area. The seed funding available to successful applicants is coming from other venture capitalists and tenants, rather than Alexandria's profits.[43]

The contrast between the financial liquidity, assets, and power of Alexandria Real Estate and its potential start-up tenants is extreme. The biotechnology industry brought in USD 2.3 billion in venture capital investments during the second quarter of 2016—a 32 percent increase over the prior quarter.[44] But given the rapidity with which smaller companies and start-ups enter and leave the sector, it is biotech real estate that proves to be the better long-term investment. With most biotech research companies being "pre-revenue" and thus unable to invest in their own buildings, or even in improvements, property with the right services and systems for research is in demand. Lease terms are typically seven to ten years, longer than in any other industry. In 2016, *CNBC News* reported that rents for laboratory space in the Kendall Square area of Cambridge, Massachusetts, had tripled in the space of twelve years. They had reached USD 71 per square foot.[45] The vacancy rate in the area had dropped to 3 percent,[46] while vacancy rates in the top-ten biotech clusters in the United States were just over 5 percent in 2015.[47] The *New York Times* noted the boom in Biotech in Cambridge had seen fashionable coffee shops, such as "Voltage Coffee" and "Art," appear on a site that had sat vacant since the razing of factories for an unrealized NASA facility.[48]

To minimize risk, scientists prefer to be in developments where there are multiple tenants with leases expiring at different times, opening the possibility of future expansion. These factors have led to companies such as Alexandria and BioMed Realty gaining a near oligopoly in the commercial laboratory sector in North America. By 2005, before building the New York campus, Alexandria owned and operated 127 properties comprising more than 8.2 million square feet (762,000 square meters) of office and laboratory space. As of June 30, 2016, the company boasted a total market capitalization of USD 12.4 billion and 18.8 million square feet (1.75 million square meters) of operating properties and development projects. While its most recent development, i3 in the University Towne Center area of San Diego, was still under construction, BioMed bought two neighboring buildings, originally developed by Nexus—they already owned the building between these two, the one occupied by Craig Venter's Synthetic Genomics and Human Longevity (figure 12.7). BioMed senior vice president commented on

Figure 12.7
Aerial View of i3 (2017) in San Diego's University Towne Center, California, with its central courtyard. Photograph by Ethan Rohloff. Courtesy of Perkins & Will.

the purchase, "It's a perfect situation, an asset with income"; further, "we have an ability to use it in a chess game in a tight market like we're seeing. It's a valuable 'hold.'"[49] San Diego county is the third largest biotech hub in the United States, behind Boston/ Cambridge and the San Francisco Bay Area. BioMed worked with the architects, Perkins and Will, on a design that boasts a campus layout around a courtyard with a performance stage, bocce ball court, herb garden, restaurant, café, and fitness area. The campus was leased to Illumina, a DNA genome sequencing provider. Blackstone Group, an investment firm founded in 1985, bought BioMed Realty in a USD 8-billion transaction that closed in January 2015. It would seem that biosciences real estate companies are playing Monopoly while selling Connect Four.

The Life of the Scientist

While governments are in large part responsible for STEM initiatives that establish scientific or "innovation" zones in their respective jurisdictions, state patronage for the actual research projects conducted within them is increasingly withdrawn in favor of market mechanisms. We have seen how those market mechanisms advantage biosciences realtors. As Lazzarato notes, "The machines of expression and constitution of the sensible (desires and beliefs) do not only act within the organization of production but also in finance."[50] Against the immense financial resources and transactions made in the industry, we must pitch the scientist whose "lifestyle" is being expressed in its promotional literature.

In *Be Creative* (2016), Angela McRobbie is sensitive to the ambivalence and ambiguity experienced by workers in the culture and creative industries. McRobbie observes, "What starts as an inner desire for rewarding work is retranslated into a set of techniques for conducting oneself in the uncertain world of creative labor."[51] The promise of pleasure in work sees the narrative of romance projected into work fulfillment. Much the same can be observed of postdoctoral students and young scientists in the life sciences, whose desire for discovery and knowledge is turned into an instrument for competition, self-discipline, and deferred reward. Young scientists and technicians are as precarious as their counterparts in the creative industries in that they, too, "are working long hours under the shadow of unemployment in a domain of intensive under-employment, and self-activated work."[52]

It is not just entrepreneurs such as Tandon who are vulnerable to income uncertainty and the potential failure of their discoveries to thrive in the marketplace. The newly social scientific organization asks each of its employees to behave like entrepreneurs, a situation reinforced by the project-based nature of the contemporary research endeavor. Research careers in the biosciences have become precarious. Biosciences research is managed in the form of temporally limited, competitively won "projects"—open and untargeted research is almost unthinkable. The project-intensive culture of science is not just a form of organization, but also a form of socialization. It shifts risk from the organizational level to the individual researchers, requiring them to develop career trajectories independent of parent organizations. At the Craig Venter Institute in La Jolla, all contracts are one year in duration, although many of Venter's partners and employers have endured this insecurity of tenure for decades. Harlem BioSpace, too, limits the tenure of its companies, requiring six-monthly applications and progress reports with the hope that "your company will succeed and graduate from the incubator within 3 years to move into independent or larger space in New York City."[53] EpiBone "graduated" to the more commodious, but far less attractive, Downstate Biotechnology Incubator in Brooklyn. The precariousness of the tenants of life sciences real estate corresponds with a post-Fordist subjectivized form of labor, which relies on an institutionalization of self-motivation, creativity, and communication skills. For those who are gregarious, it improves working conditions through less hierarchical social organization, empowers through self-government, and achieves a greater sense of belonging to a collective and collaborative group. Certainly, this is the belief that colors the rhetoric around laboratory design, and given the richness of facilities, it is easy to imagine pleasurable and convivial scientific working lives. Yet, it exposes scientists to funding uncertainty and exploits and produces a precarious condition for workers. Effectively, its subjects are drawn into self-governance.

And while images of collaboration abound, this self-governance creates new divisions between and hierarchies of science workers. A scientific intelligentsia armed with instrumental, market-based values and expertise are finding themselves in positions of increased power where they can better negotiate their conditions and careers than the manual and technical staff that support them. Business theorists refer to the workers in any industry who can move from project to project, as "intellectual mercenaries." Tammy Erickson argues that to succeed in this context workers need to build networks, develop a "personal brand," and engage in continuous education.[54] Further, "these tendencies represent a critical shift in the nature of class relations in contemporary capitalism, with cultural 'assets' providing key resources that distinguish newly ascendant strata."[55] New social skills—the ability to attract and cooperate with team members, initiate and maintain conversation with peers and superiors, and persuade others of the value of your work—are required, along with newly fit and healthy bodies honed in on-site gymnasia. In this light, the restaurants, yoga classes, and foosball tables of the new research laboratory, and the dynamic urban contexts in which they are increasingly situated, are not there merely to entice the ingredients of innovation. They have ongoing work to do in maintaining "fitness." "Diddling around" takes on a new meaning.

Michel Foucault noted the ease with which individuals succumb to those forces that repress them, while Deleuze and Guattari spoke of the pleasure of being at one with the gears of the machine.[56] Or as Alan Liu succinctly expresses the situation, "Design is how we can be dominated by instrumental rationality and love it, too."[57]

13 Conclusion

In the summer of 2013, we visited Cambridge University's 2012 Royal Institute of British Architects Stirling Prize–winning Sainsbury Laboratory (2011). Afterward we wandered down the main street of Cambridge to the Fitzwilliam Museum for cultural respite. The museum was founded in 1816 to display the vast collections of Richard Fitzwilliam, 7th Viscount Fitzwilliam of Merrion. Its patron was a musical antiquarian, a "grand tourist," and what we call today "a collector." Fitzwilliam bequeathed to the university his extensive library and a vast collection of works of art, medieval and Renaissance manuscripts, and autographed musical notation, along with the funds to construct a grand classical building to house them. The collection, albeit focused on cultural artifacts, suggests a penchant for the natural history of the time. This was the period of Erasmus Darwin's forays into evolution, Jean-Baptiste Lamarck's investigations of invertebrates and transgenerational inherited traits, and Etienne Geoffroy Saint-Hilaire's explorations of monstrosity. This collection was as vast as the animals Comte de Buffon had recorded, the fossil collections Georges Cuvier obsessed over, and the skeletons Richard Owen coveted.

The displays of world culture at the Fitzwilliam Museum are organized into patterns, which echo the arrays of plants and animals in botanical and natural history museums. The museological underpinnings acknowledge timeframes, geographical distribution, and morphological consistencies. On the day we visited, the exhibition "Origins of the Afro Comb: 6000 Years of Culture, Politics, and Identity" was open. It was curated by Senior Assistant Keeper of Antiquities Sally-Ann Ashton, an archaeologist. In the Mellon gallery were thousands of objects that tracked the historical progression of grooming from one side of Africa to another and in the diaspora. The key objects of the exhibition, displayed side by side, were a 5,500-year-old bone comb, excavated in 1898 from a cemetery at Abydos, and a plastic Black Fist comb, one of thousands mass produced in the 1970s. Between these two geohistorical artifacts, hundreds of combs were arranged to suggest geographic migrations and historical developments (figure 13.1). When

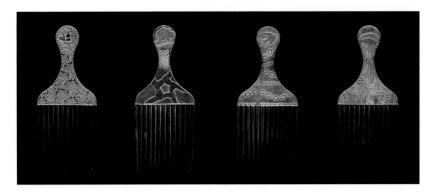

Figure 13.1
Four of ten silver-plated combs made by Russell Newell in 2013, which tell the story of Jamaica's Accompong Maroons, part of the permanent collection at the Fitzwilliam Museum, Cambridge, UK, included in the Afro Comb exhibition, Fitzwilliam Museum Cambridge. Photograph by Social Creatures.

viewed across this scale, every single Afro comb appeared to fall into lines: chronological, geographical, and morphological. The combs varied from roughly carved to refined implements, in wood, palm leaf, and bamboo to ivory, bronze, and plastic. There were shifts in the richness and references in ornamentation of the combs. Some had images of animals; others had anthropomorphic images, tiny geometric patterns, and images of an iconic clenched fist. There were periods when the implement changed quickly and others when it changed little. These developments led us to speculate upon how hair too may have changed, genetically, structurally, stylistically, and semiotically.

The time period covered by this exhibition was about 6,000 years. Every aberration seems negligible in the scheme of things. The combs were presented, as one critic noted, as if there "were some sort of evolutionary process that began in ancient Egypt and reached its apotheosis in New York in the 1970s. That is clearly wrong in all sorts of ways. Rather, there are hundreds and thousands of histories."[1] Indeed, the fate of many of the aberrations of this exhibition was a display case of combs with "lost histories." These combs were of uncertain origin. We imagine also the uncertainty relates to their differences. They just don't seem to fit comfortably into the narrative of the Afro comb and are thus deemed to be "lost." These aberrations were not the only ones, though. Like Charles Darwin's isolated island species, oddities also appeared in the main display. These oddities were productive monsters, marking moments of bifurcation in the narrative of the Afro comb—moments when there seems to have been proliferations of competing forms, indulgences in radically altered ornamentation or material usage, or accelerations of development. These oddities spawned new and innovative combs, marking moments when the comb went toward what might be called a brush or when it might function as a weapon or as an affirmation of Black Power in the context of racial oppression—moments when the comb would pass through more than just the hair on one's head. The images complementing the display were of the hair to which the combs responded and the hairstyles and political communities they helped construct over the six millennia.

In contrast, our consideration of the laboratory covers only the first two decades of the twenty-first century. The perspective is chronologically tight, geographically vast, and morphologically perverse. We note in this period an immense flourishing. The laboratories we have seen and explored range in scale from the single room start-up incubator at Harlem Biospace through to the massive town-scale investments of Janelia Farm and Cold Spring Harbor Laboratory (CSHL). Their design aesthetic varies from the austere Fabrikstraße 22 to the shimmering, jewel-like NavarraBioMed. Materially, they encompass the technically sophisticated glazed skin of the South Australian Health and Medical Research Institute (SAHMRI) and the quaint timbers and shingles of the CSHL. The selection stretches from the remote and isolated Biosphere 2 to the urban real estate of Alexandria's project in Manhattan and includes the educational program of the Dr. Seuss–like Blizard Building along with the clinical investments of the Centre for the Unknown. Some of the laboratories suggest intimacy, yet are just one element in a larger network of interests, as was the case of Craig Venter's institute. Others construct communities—stable and identifiable; mobile and fun; ambitious and reified; and where animals are bred for sacrifice. These laboratories have sprung up across the globe, from one country to another, from one economy to another. It has been less a diaspora than a rhizomatic emergence.

Were we to lay our laboratories out as one might lay out Afro combs, there would be a thousand ways to organize the display. Indeed, the organization of this book was contentiously debated. Following the chronologies, geographies, and morphologies that constituted the Fitzwilliam museum's display seemed inappropriate. Our chronology was collapsed into two decades. Given the fluid global economies in which the biosciences are situated and the mobility of architects and scientists, geographic positioning seems to be beside the point. At first glance, the morphologies of these laboratories give little clue to any overarching trajectory or ordering principle. We suspect if the curator of the museum exhibition had been invited to organize our laboratories, there is a possibility they may have all ended up in the "lost histories" display case. We looked for a single line of investment to which all our bioscience laboratories aspire and hoped to locate something like the "Gothic line" to which all cathedrals aspired: a single line to heaven that gave order to all below it. Such a line is elusive in the complex proliferations of investment that come to constitute the contemporary laboratory. The perspectives we have on life today are complicated and multifarious. In an era of poststructuralism, speculative realism, neo-materialism, and the Anthropocene the question of meaning seems moot. These uncertainties are not the preserve of philosophers; architects, too, have become jittery about meaning-making. Indeed, even within one building, architects themselves struggle: Renzo Piano, for example, first declares his suspicion of metaphors before describing the Jerome L. Greene Science Center (2016) for neuroscience research in Harlem, New York, in both metaphoric and performative terms, as "a palace, it is a palace of light" and as "a factory exploring the secret of the mind" (figure 13.2).[2]

The year following the Cambridge visit, and after many mojitos on the terrace of Darwin's Café in Lisbon, we decided to hedge our bets and began to organize this book around a tension. This tension relates to the dual sense we have that laboratories are constructed and that they also construct. We looked to the key moments in laboratories where this tension plays itself out most intensely and came to fixate upon the *boundaries* that come to constitute science and the laboratory; the *expression* of scientific ideas

Figure 13.2
Jerome L. Greene Science Center in New York City, under construction in 2015. Architect: Renzo Piano. Photograph by author.

and ideals, and the socio-construction, or what we call the *socialization*, of the contemporary scientist. Our bet paid dividends—not because these categories generated singular lines of thought that led to simple conclusions, but because they, too, spoke of the complex and perplexing problem of summing anything up in simple terms. The very boundaries that constituted science, the demarcations of biomedical laboratories and their areas of physical containment, had bled. The expressions of science came to be problematic and, indeed, moments of expression floating free of content, vague and ephemeral. Our sociocultural constructions of the scientist unfolded into networks, spreadsheets, property borders, and the sliding pages of social media. The motto of Jules Verne's Nautilus, "Mobilis in mobile," now stands for all our attempts to pinpoint and organize.

This is not to suggest we would not recognize a laboratory, an expression of science, or a scientist. They are as identifiable as Afro combs, despite their diversity. What the motilities of the contemporary biosciences laboratory suggest is that there is no archetypal architecture, no *urtypen* or *idéale*, at this moment. There is no definitive concept, form, material, organization, or aesthetic to which all laboratories appeal. However, odd *vestigial* organs in the laboratories and points of accretion, which may or may not have functional value, are recognizable. The plaza of the Salk Institute is one such vestigial organ, which recurs in almost all the laboratories we considered. It is sometimes heavily pronounced and clearly articulated, as for example, in the plazas open to the sky of the Centre for the Unknown in Lisbon and the reflective roof pond of the Barcelona Biomedical Research Park (PRBB). It is less pronounced in the mews of the Blizard Building or Novartis's Fabrikstraße 22. It is hidden deep within the flesh of some of our laboratories, like the bone structure of the blue whale's fins. The shrunken courtyard of the J. Craig Venter Institute (JCVI) and the geological caverns of the Pharmaceutical Science Building (PSB) seem to be particular consumptions or inversions of the organ.

We also encountered across the majority of our laboratories what are best described as *vestiges of ideas*, rather than architectural elements. These were the rhetorical drives that occurred as the default refrains of the contemporary laboratory. Statements that related to essential notions of health and wellbeing, the future, investment, and impromptu conversation, as well as the formation of communities to engage with the challenges of the scientific endeavor. Statements related to these ideas were repeated *ad nauseum* and came to occupy our dreams like tunes you can't get out of your head. Such rhetorical acts operated like liturgy circulating the globe to give a sense of order to the complexity and dynamisms of life. The consistency of the rhetoric that architects, architectural critics, building managers, scientists, and technicians brought to the laboratory belies the rich proliferations and indulgences of the architecture.

While we observe the consistency of rhetoric, in each chapter we zoom in on single laboratories, bringing to bear upon them our architectural analysis. We apprehend them first as material artifacts. We are more precise than was Bruno Latour in naming and describing the architectures of each laboratory: the volume and massing of form; the way materials are assembled and detailed; the choreography of spatial arrangements; the shape, transparency, and shading of windows; the species of plant in gardens; the atmospheric effects of light, heat, and chilled air-conditioning. It is only through forensic investigation that we are able to reveal how *this* particular assemblage of boundaries, images, spatial organization, and accompanying rhetoric has found expression as architecture. Thus, we chart the forces that are generic (formal laws and regulations, institutional interests, state ideologies, cultural taboos, and general social and political conditions) and those forces that are contextual (constraints set by clients, budgets, scientific concerns, the specifics of a particular site, and the involvement of other building specialists). While each laboratory responds to a common set of forces, we have found that each is also highly particularized.

But we do more than relating architecture to context. The laboratory is more than a form of condensation of expression for forces upon it. Georges Canguilhem suggests science is not a matter of fact or fiction, but rather, a matter of being operative or not: a scientist may *dire vrai* (speak the truth) but not be *dans le vrai* (with the true) of contemporary scientific discourse.[3] What he is suggesting is that a scientific fact is regarded as a truth only when it resonates with the forces of scientific discourse and theorization of the time. That is, both the fact and the context are involved in a type of dance that produces either discord or harmony. The relation of the laboratory to science is similar. It is at once the product and producer of science. That is to say, the laboratory as artifact is operative. It intervenes. It shapes its subject, science, as it is shaped by it. It constructs scientists, just as scientists construct it.

Latour described a similar phenomenon under the term *factish*. Extending his proposition, we might coin the term *artifactish* to describe the laboratory. What makes the laboratory artifactish is the resonance between the architectural object and expression and the fictions or rhetorical structures of science. The laboratory propels science forward, it gives succor, it amplifies. Janelia Farm's closed world is a vehicle for "seeding" and "harvesting" promising scientists, translating economic excess into intellectual capital and back again into economic gain. The Blizard borrows the floating florid forms of childhood (or imagined childhood) in order to repurpose the working laboratory as an educational and expository tool. CSHL, meanwhile, holds up a mirror to its local philanthropists, offering them a cause and a venue for their excessive accumulation of wealth and, in turn, producing discoveries by which privileged bodies might be maintained and lives prolonged.

We imagine zooming out and seeing the long history of the architecture of the biosciences laid before us in a display like that at the Fitzwilliam Museum. The display might start with the birth of biology as a discipline. If one were to zoom into the two-decade period at the turn of the twenty-first century, it would surely reveal itself to be a richly fertile moment—a moment when the laboratory found itself central to an ecology of forces that caused it to reproduce at an extraordinary rate and come to occupy every corner of the globe; when its formal expressions, its architecture, and its ornamentation blossomed and inspired. From this moment the laboratory might fragment further or dissolve into other typologies. It might transition in its attachment to translational medicine, the clinic, and the hospital, or be reabsorbed into a new formation. The sites of scientific research might be swallowed into the spreadsheets and behemoth economies of Big Pharma or proliferate like a coral into thousands of kitchen labs and maker-spaces for biohackers and hobbyists. It may join the country clubs of Long Island or find its terraces washed into the ocean as a result of climate change. This select period has been, for the laboratory, what the thirteenth and fourteenth centuries were for the cathedral, or indeed what the eighteenth and nineteenth centuries were for the museum—a period of immense change and mass investment.

Cultural artefacts like architecture tend not to evolve gradually but to develop in fits and starts. We would like to imagine this is a key difference between the chronologies, geographies, and morphologies of cultural artifacts and the evolution of organisms. Stephen Jay Gould and Niles Eldredge, the authors of the theory of "punctuated equilibrium," think otherwise.[4] Species, they believe, are particularly stable for long periods and then, given the right circumstances—competition and pressure, abundance of resource or strategic isolation—they mutate quickly and sometimes with incredible diversity. In 2002, Eldredge tested the relevance of the patterns of evolution to the histories of cultural artifacts upon his collection of 500 or so cornets—a smaller version of the trumpet. He identified seventeen characteristics of the cornet's anatomy, in the same manner that he had categorized trilobites in his day job as curator in the Invertebrate Paleontology Department at the American Museum of Natural History. He fed the data through the computer program he used to organize and map the evolution of natural specimens.[5] Eldredge describes it as "a mapping exercise of the distribution of characteristics" among a series of objects.[6] He observed that the lines in the cornet evolutionary tree were thoroughly confused. Biological evolution is characterized by a cone of diversity. Change occurs as a series of transformations and modifications of past forms because organic beings are dependent on the vertical transmission of inherited traits, from parents to offspring. If a species dies out, it cannot reappear. This is not the case for cornets, with innovations appearing spontaneously, sometimes with no physical precursor. Horizontal transfer was rife, as makers copied each other. Outdated instruments reappeared with new designs, or borrowed from across branches. There was no extinction but an evolutionary dynamic in which the rules of inheritance were not based on Mendelian genetics. There was for cornets, too, periods of stasis or stability, and periods of rampant experimentation. Every now and again the cornet made leaps, sometimes geographic, morphological, and, we imagine, in the type of sound it played. Indeed, sometimes the form of the music seemed to construct the cornet. The cultural artifact can do this in a manner that an organism cannot. It can reach backward as it thrusts itself forwards. It can construct (just as it is constructed). We could play a piece by G. F. Handel on a cornet manufactured for jazz on a terrace of a biosciences laboratory café (or at least, we could if we knew how to play and if only we released our grip on the mojitos).

For bioscience laboratories ideas and forms are circulated, copied, abandoned, and sometimes revisited. The glass vaults and biomes of Biosphere 2, for example, make an unexpected return in Moshe Safdie's proposal for the Albert Einstein Medical School (due for completion in 2021) in Sao Paolo, Brazil (figure 13.3). The island motif recurs in the circular campus, 280 meters in diameter, that is Herzog and de Meuron's Skolkova Institute of Science and Technology (2018) in Russia. Here, what is

Figure 13.3
Architect's rendering of the atria of the Albert Einstein Education and Research Center (2021), Sao Paulo, Brazil. Architect: Safdie Architects, courtesy of Safdie Architects.

described by the architects as one of five "urban islands, inserted into the picturesque landscape" is charged with creating a self-sufficient, bounded, scientific community while, in seeming contradiction, distancing itself politically and morphologically from the Closed Cities of the Soviet era.[7] Despite its monumental and castellated appearance its architects maintain that "its courtyards are open and accessible to the public" and that the building is permeable (figure 13.4). Encircled by a vehicular moat and located 30 kilometers from the center of Moscow, it seems unlikely that any "public" are going to wander into this complex.[8] The détente between local heritage and scientific research ambitions that we saw at CSHL recurs in the gymnastics performed by the Wellcome Genome Campus, Hinxton, UK, to placate its neighbors and advance its 2018 masterplan for development.[9] Gensler architects have installed a room that "resembles an oversized cell," a paler version of that at the Blizard but recognizable all the same, for Johnson & Johnson's research incubator "JLabs" in Houston (2015) and Toronto (2016). It is not impossible that the modernist laboratory—or at least its architectural predilections—could make a reappearance. It is apparent in our study that the impulse for change has not been confined to the opportunities that architects have to create, adapt, or share ideas. In this, it is not enough to dismiss the diversity

Figure 13.4
Skolkovo Institute of Science and Technology (2018), Moscow Oblast, Russia. Architect: Herzog & de Meuron Architects. Photograph by Iwan Baan.

Figure 13.5
Atrium of the Victorian Comprehensive Cancer Center (VCCC) (2016), Melbourne, Australia.
Architects: Silver Thomas Hanley, DesignInc and McBride Charles Ryan. Photograph by authors.

of laboratory architecture as generalizable to the plurality of architecture in the twentieth century. Indeed, we would argue that such plurality, the very same that Vittorio Lampugnani decried in museums and sought to stifle at Novartis, is a specifically architectural narrative. It is one that overstates the agency of the architect at the same time as it downplays the capacity for architectural expression to embody, knowingly or not, external forces and mechanisms. Architectural plurality feeds on and responds to the transformation of institutions and disciplines. The museum, too, would benefit from a close study of its artifactish ecology akin to that we have attempted here.

The rapid changes and diversity of form in the laboratory have been in part a response to the intense desires among scientists and public and private investors for discovery. As Venter said in relation to energy, the nation that discovers an alternative fuel source *owns the future*. This is bold rhetoric, but the idea of owning the future—be that through curing cancer or achieving human "'longevity"'—is pervasive. The last laboratory we visited was the Victorian Comprehensive Cancer Center (VCCC) (2016) in Melbourne, which, according to both architect and client, expresses a narrative of "hope." Designed by Silver Thomas Hanley, DesignInc, and McBride Charles Ryan architects, the VCCC is a seven-story cancer diagnosis and treatment clinic below a five-story laboratory building. In this context, hope refers to death deferred and pain alleviated, but it might equally refer to the professional ambitions of its scientists—to win the race to the cure. Hope is a frequently repeated refrain. In place of the faith that animated the cathedral and drew the architectural trajectory upward, hope, we are told, is expressed in the spiraling forms that circle this building's façade and carve out a dizzying, elliptical atrium inside (figure 13.5).

Whereas an oratory is a small chapel for private prayer, hope seems to require a larger space. But it remains as fickle as a hedge fund, and, in the case of the biosciences, it is perhaps a wondrous delusion: a grasping for immortality. If hope is the emotional driver of the biosciences laboratory it is no wonder the stories told by its architecture are complex and mobile in their references and trajectories, historicides and reiterations. We are busy constructing laboratories as they construct stories about forms of discovery, science, and scientists, and about life. This is why we have called this book *LabOratory*.

Notes

1 Introduction

1. Many scholars have considered the status of the sciences as the major knowledge-producing social institutions in secular society. See for example: C. E. Kenneth Mees, "The Production of Scientific Knowledge," *Science* 46, no. 1196 (November 30, 1917): 519–528; Richard Whitley, *The Intellectual and Social Organization of the Sciences* (Oxford: Clarendon Press, 1984); Stephen Hillgartner, Clark A. Miller, and Rob Hagendijk, eds., *Science and Democracy: Making Knowledge and Making Power in the Biosciences and Beyond* (London: Routledge, 2015).

2. "Revenue of the Worldwide Pharmaceutical Market from 2001 to 2017 (in billion U.S. dollars)," The Statistics Portal, accessed March 20, 2019, https://www.statista.com/statistics/263102/pharmaceutical-market-worldwide-revenue-since-2001/.

3. Oliver Wainwright, "Why the Sainsbury Laboratory Deserved to Win the Stirling Prize," *Guardian*, October 15, 2012, http://www.guardian.co.uk/artanddesign/2012/oct/15/sainsbury-laboratory-deserved-stirling-prize.

4. "The Sainsbury Laboratory in Cambridge by Stanton Williams Wins the RIBA Stirling Prize 2012," Royal Institute of British Architects (RIBA), accessed October 3, 2015, https://www.architecture.com/RIBA/Contactus/NewsAndPress/NewsArchive2002-2012/AwardsNews/News/2012/TheSainsburyLaboratoryinCambridgebyStantonWilliamswinstheRIBAStirlingPrize2012.aspx.

5. For a detailed discussion of the spaces of early science, see David Livingstone, *Putting Science in Its Place: Geographies of Scientific Knowledge*, (Chicago: University of Chicago Press, 2003), 1–86.

6. Steven Shapin, *A Social History of Truth: Civility and Science in Seventeenth-Century England* (Chicago: University of Chicago Press, 1994), 409.

7. Bruno Latour, "The Costly Ghastly Kitchen," in *The Laboratory Revolution in Medicine*, ed. Andrew Cunningham and Perry Williams (Cambridge: Cambridge University Press, 1992), 299.

8. Latour, "The Costly Ghastly Kitchen," 299.

9. Sven Dierig, "Engines for Experiment: Laboratory Revolution and Industrial Labor in the Nineteenth-Century City," *Osiris* 18 (2003): 118.

10. Oswald Grube, "The Birth of the Modern Research Building in the USA," in *Principles of Research and Technology Buildings*, ed. Hardo Braun and Dieter Grömling (Basel: Birkhauser, 2005), 21.

11. Grube, "The Birth of the Modern Research Building in the USA," 21.

12. "Rudolph in the Movies—Brainstorm (1983)," Paul Rudolph Foundation News (June 26, 1983), accessed November 19, 2018, http://paulrudolph.blogspot.com/2009/06/rudolph-in-movies-brainstorm-1983_26.html.

13. Bruno Latour, "From the World of Science to the World of Research?" *Science* 280, no. 5361 (1998): 208.

14. Bruno Latour, "To Modernize or to Ecologize? That's the Question," in *Remaking Reality: Nature at the Millennium*, ed. N. Castree and B. Willems-Brawn, (New York: Routledge, 1997), 232.

15. Theatrical performances such as Michael Faraday's Friday evening presentations to Victorian audiences concealed the untidiness of experimentation behind smooth artistry, just as today's exhibition spaces and laboratories, websites and media releases present a polished and curated picture of experimentation. See Livingstone, *Putting Science in Its Place*, 23–25.

16. KMD monograph (March 1, 2016): 12, https://issuu.com/kmdarchitects/docs/kmd_monograph_02-29-16_sf_low_res.

17. Nigel Thrift, "Re-Inventing Invention: New Tendencies in Capitalist Commodification," *Economy and Society* 35, no. 2 (2006): 294.

18. In 2013 Botswana commenced construction on a massive 270,000 square feet Innovation Hub in Gaborone, designed by SHOP Architects, which brings science, medicine, engineering, and business innovation together. For a fascinating account of Bolivia's attempts to develop a science economy. see Katherine McGurn Centellas, "For Love of Land and Laboratory: Nation-Building and Bioscience in Bolivia," PhD diss., University of Chicago, 2008.

19. "About Weill Hall," Cornell University homepage, accessed November 22, 2018, http://blogs.cornell.edu/whfs/about-weill-hall/.

20. Ewen Callaway, "Europe's Superlab: Sir Paul's Cathedral," *Nature* 522, no. 7557 (June 23, 2015): 406–408, http://www.nature.com/news/europe-s-superlab-sir-paul-s-cathedral-1.17827.

21. Sir Paul Nurse, quoted in Nicola Davis, "Francis Crick's 700m Altar to Biomedical Science," *Guardian*, February 17, 2016, https://www.theguardian.com/science/2016/feb/17/francis-crick-institute-biomedical-science-research.

22. HOK News, "*Nature* Explores Design of Francis Crick Institute, London's Collaborative Super Lab," January 15, 2016, http://www.hok.com/about/news/2016/02/15/nature-explores-design-of-francis-crick-institute-londons-collaborative-super-lab/.

23. The Francis Crick Institute was nominated for the Carbuncle Cup, the UK's award for the worst building of the year. Architecture critic Oliver Wainwright described it as a "dog's dinner" and likened it to "a portly scientist stuffed into an ill-fitting suit and crowned with a shouty hat."

Oliver Wainwright, "Francis Crick Institute: Cathedral of Science 'Looks Better from 1,000ft,'" *Guardian*, September 3, 2016, https://www.theguardian.com/artanddesign/2016/sep/02/francis -crick-institute-review-it-looks-better-from-1000-feet.

24. Karin Knorr Cetina, "The Couch, the Cathedral, and the Laboratory: On the Relationship between Experiment and Laboratory in Science," in *Science as Practice and Culture*, ed. Andrew Pickering (Chicago: University of Chicago Press, 1992), 128.

25. Gutman's research into building satisfaction in the 1970s began with Louis Kahn's Richards Laboratory Building at the University of Pennsylvania medical school. The building was acclaimed by the architectural profession, but condemned in scientific circles. One scientist suggested to Gutman that the building had "seriously impeded the progress of medical science" (Robert Gutman and Barbara Westergaard, "Building Evaluation, User Satisfaction, and Design" in *Designing for Human Behaviour: Architecture and the Behavioural Sciences*, ed. Jon T. Lang, Charles Burnette, Walter Moleski, and David Vachon [Stroudsburg, PA: Dowden, Hutchinson and Ross, 1974], 327, 324).

26. Frank Gehry and Nancy E. Joyce, *Building Stata: The Design and Construction of the Frank O. Gehry's Stata Center at MIT* (Cambridge, MA: MIT Press, 2004), xv.

27. Spencer Reiss, "Frank Gehry's Geek Palace," *Wired Magazine* 12, no. 5 (May 2004), http:// www.wired.com/wired/archive/12.05/mit.html.

28. It is impossible to longitudinally and accurately measure changes in research productivity, let alone the significance of that research or the degree of cross-disciplinarity or innovation. Measurements of research outcomes are often unreliable and contested; they are not like counting the number of telephone relays assembled in a day (as was the case in the infamous Hawthorne experiments of 1924–1935, which demonstrated that workers worked harder simply because they were under scrutiny). We would need a very long tracking of the same individuals using the same measurement criteria, yet in the face of dramatic changes to funding and accountability for universities across Europe, the UK, North America, and Australasia in the past two decades, research productivity measures have changed too greatly for longitudinal comparison. Furthermore, many of the research facilities constructed in the last decade are for newly constituted interdisciplinary and cross-institutional groups and bring researchers and academics together for the first time working on essentially new projects. They are novel organizations that develop their workplace culture in concert with their occupation of new premises. Even if co-location were proven to make a difference, it would not establish a case for the influence of architectural settings. Furthermore, new buildings are typically prompted by growth in successful organizations, and the building then confirms that growth by underscoring the success of the group. Lastly, the sites of new knowledge and transfer are not bound by the precise location of laboratory practices. In fact, the sites for transferring scientific knowledge extend to include the campus and the city, conference venues and hotels, convention centers where trade shows are staged, journals, and social networking sites. The insights that Frank Gehry imagines will take place because the new building may well occur in any location or social context.

29. Jeanne Kisacky, "History and Science: Julien-David Leroy's 'Dualistic Method of Architectural History,'" *Journal of the Society of Architectural Historians* 60, no. 3 (2001): 260–289.

30. Viollet-le-Duc, quoted in A. T. Edwards, *Architectural Style* (London, 1926), republished as *Style and Composition in Architecture* (London, 1944), cited in Philip Steadman, *The Evolution of Designs: Biological Analogy in Architecture and the Applied Arts* (Abingdon, UK: Routledge, 2008), 69.

31. Georgina Ferry, "The Art of Laboratory Design," *Nature* 457, no. 541 (January 29, 2009): 1, http://www.nature.com/nature/journal/v457/n7229/full/457541a/html.

32. Daniel Watch, Deepa Tolat and Gary McNay, "Academic Laboratory," *Whole Building Design Guide* (Washington DC: National Institute of Building Sciences, 2010), 4, http://www.wbdg.org /design/academic_lab.php.

33. Shapin, *A Social History of Truth*, 410.

34. Thrift, "Re-Inventing Invention," 293.

35. Architect Will Alsop quoted in a media release, published on Arcspace.com, November 21, 2005. http://www.arcspace.com/features/alsop-architects/blizard-building/.

36. Chris L. Smith and Sandra Kaji-O'Grady,"Exaptive Translations between Biology and Architecture," *Architecture Research Quarterly* 18, no. 2 (2014): 155–166.

37. Robert Kohler, "Labscapes: Naturalizing the Lab," *History of Science* 40, no. 4 (2002): 473.

2   Oratory in the Lab

1. We are mindful that the word play in our book title is not original. Lab.Oratory, a men-only sex club in Berlin with industrial decor and an overheated atmosphere, claimed the moniker long before we did; http://www.lab-oratory.de/.

2. Kim Nasmyth, quoted in University of Oxford News, "New Biochemistry Building Opens," accessed March 27, 2012, http://www.ox.ac.uk/media/news_stories/2008/081212_1.html.

3. Jean-Francois Lyotard, *The Postmodern Condition: A Report on Knowledge*, trans. Geoff Bennington and Brian Massumi (Manchester, UK: Manchester University Press, 1984), 45.

4. Henning Schmidgen, *Bruno Latour in Pieces*, trans. Gloria Custance (New York: Fordham University Press, 2014), 29, 33.

5. Bruno Latour and Steve Woolgar, *Laboratory Life: The Construction of Scientific Facts* (Princeton, NJ: Princeton University Press, 1986), 41.

6. Latour and Woolgar, *Laboratory Life*, 48.

7. Latour and Woolgar, *Laboratory Life*, 51.

8. Latour and Woolgar, *Laboratory Life*, 52.

9. Latour and Woolgar, *Laboratory Life*, 87.

10. Latour and Woolgar, *Laboratory Life*, 87.

11. Latour and Woolgar, *Laboratory Life*, 243.

12. During the science wars of the 1990s, Latour's insistence on the historical contingency and subjectivity of science came under criticism from biologist Paul Gross and mathematician Norman Levitt, philosopher John Searle and mathematician and physicist Alan Sokal. See Paul Gross, Norman Levitt, N. and Martin Lewis, eds., *The Flight from Science and Reason* (New York: New York Academy of Sciences, 1996); John Searle, *The Construction of Social Reality* (New York: Free Press, 1995); Alan Sokal and Jean Bricmont, *Fashionable Nonsense: Postmodern Abuse of Science* (New York: Picador, 1999), first published as *Impostures intellectuelles* by Editions Odile Jacob in 1997; Alan Sokal M. Lynch, and S. Cole, "Science and Technology Studies on Trial: Dilemmas of Expertise," *Social Studies of Science* 35, no. 3, (2005): 269–311.

13. Paul Gross and Norman Levitt, *Higher Superstition: The Academic Left and Its Quarrels with Science* (Baltimore, MD: John Hopkins University Press, 1998), 60.

14. Latour and Woolgar, *Laboratory Life*, 32.

15. Latour and Woolgar, *Laboratory Life*, 19.

16. Latour and Woolgar, *Laboratory Life*, 19.

17. Latour and Woolgar, *Laboratory Life*, 320, fn. 3.

18. Paolo Fabbri, personal communication with Henning Schmidgen, January 13, 2011, quoted in Schmidgen, *Bruno Latour in Pieces*, 36.

19. Latour published a paper with Fabbri in 1977 that portrays this agonistic dimension of scientific work in the laboratory in *Actes de la Recherche en Sciences Sociales*, the journal founded by Pierre Bourdieu. Bruno Latour and Paolo Fabbri, "La Rhétorique de la Science Pouvoir et Devoir dans un Article de Science Exacte," *Actes de la Recherche en Sciences Sociales* 13 (1977): 81–95.

20. Bruno Latour and Paolo Fabbri, "The Rhetoric of Science: Authority and Duty in an Article from the Exact Science [1977]," trans. Sarah Cummins, *Technostyle* 16, no. 1 (2000): 115–134.

21. Latour and Fabbri, "The Rhetoric of Science," 132.

22. Latour and Fabbri, "The Rhetoric of Science," 132.

23. Jonas Salk, "Introduction" (1979), in Latour and Woolgar, *Laboratory Life*, 12.

24. Irit Rogoff, "Gossip as Testimony: A Postmodern Signature," in *Generations and Geographies in the Visual Arts: Feminist Readings*, ed. Griselda Pollock (New York: Routledge, 1996), 58.

25. Rogoff, "Gossip as Testimony," 59.

26. Gilles Deleuze and Félix Guattari, *What Is Philosophy?*, trans. Hugh Tomlinson and Graham Burchill (London: Verso, 1994), 144.

27. Deleuze and Guattari, *What Is Philosophy?*, 199.

28. Paul Feyerabend, *Against Method: Outline of an Anarchistic Theory of Knowledge* (New York: New Left Books, 1975), 308 (this is a translation of *Contre la méthode; ou Esquisse d'une théorie anarchiste de la connaissance* [Paris: Seuil, 1975]. For a summary of the contentions, see Nicholas Jardine and Marina Frasca-Spada, "Splendours and Miseries of the Science Wars," *Studies in History*

and Philosophy of Science 28, no. 2 (1997): 219–235; Philip Kitcher, "A Plea for Science Studies," in *A House Built on Sand: Exposing Postmodernist Myths about Science*, ed. Noretta Koertge (Oxford: Oxford University Press, 1998), 32–56; James Robert Brown, *Who Rules in Science?: An Opinionated Guide to the Wars* (Cambridge, MA: Harvard University Press, 2001).

29. Suzannah Lessard, *The Architect of Desire: Beauty and Danger in the Stanford White Family* (New York: Delta, 1991).

30. Schmidgen, *Bruno Latour in Pieces*, 28.

31. Latour and Fabbri, "The Rhetoric of Science," 131.

32. Luis Barragán, quoted in Louis Kahn, letter to James Britton, *Urban Design Review*, June 12, 1973; Luis I. Kahn Archives, University of Pennsylvania, Box LIK 27: "Salk Institute for Biological Studies, La Jolla, CA."

33. Quoted in Michael Crosbie, "Add and Subtract," *Progressive Architecture* 74, no. 10 (1993): 48.

34. Robert Venturi and Denise Scott Brown, "A Protest Concerning the Extension of the Salk Center," in *Iconography and Electronics upon a Generic Architecture: A View from the Drafting Room*, ed. Robert Venturi (Cambridge, MA: MIT Press, 1996), 82.

35. Paul Goldberger, "Imitation that Doesn't Flatter," *New York Times*, April 28, 1996, http://www.nytimes.com/1996/04/28/arts/architecture-view-imitation-that-doesn-t-flatter.html?pagewanted=1.

36. Karen Stein, "The Salk Institute Remains a Modernist Beacon," *Architectural Digest*, January 31, 2013, http://www.architecturaldigest.com/architecture/2013-02/louis-kahn-salk-institute-la-jolla-california-article.

37. Gross and Levitt, *Higher Superstition*, 57.

38. Quoted in Ann Gibbons, "The Salk Institute at a Crossroads," *Science* 249, no. 4967 (1990): 360.

39. Thomas F. Gieryn, "What Buildings Do," *Theory and Society* 31, no. 1 (2002): 46.

40. Our apologies to Emily Pawley, whose study of the mulberry craze provoked by agricultural improvement societies in late 1830s upstate New York is, in fact, riveting. The interaction of monetary and natural economy that she studies touches on some of the ideas around the restricted and general economy that we discuss in chapter 7. Emily Pawley, "'The Balance Sheet of Nature': Calculating the New York Farm, 1820–1860," PhD diss., University of Pennsylvania, 2009.

41. Robert E. Kohler, "Lab History: Reflections," *Isis* 99, no. 4 (2008): 761.

42. Kohler, "Lab History: Reflections," 762.

43. Kohler, "Lab History: Reflections," 764.

44. Kohler, "Lab History: Reflections," 766.

45. Goldberger, "Imitation that Doesn't Flatter."

46. Stuart Leslie, "'A Different Kind of Beauty': Scientific and Architectural Style in I. M. Pei's Mesa Laboratory and Louis Kahn's Salk Institute," *Historical Studies in the Natural Sciences* 38, no. 2 (2008): 173–221.

47. There are 1,100 life sciences and more than eighty research institutes in the San Diego region; http://www.sandiegobusiness.org/industry/lifescience.

48. Kohler, "Lab History: Reflections," 766.

49. Emile Durkheim, "Methods of Explanation and Analysis (1899)," in *Emile Durkheim: Selected Writings*, ed. Anthony Giddens (Cambridge, UK: Cambridge University Press, 1972), 88.

50. Kohler, "Lab History: Reflections," 767.

51. See Steven Shapin and Simon Schaffer, *Leviathan and the Air Pump: Hobbes, Boyle and the Experimental Life*, (Princeton, NJ: Princeton University Press, 1985); Steven Shapin, *A Social History of Truth: Civility and Science in Seventeenth-Century England* (Chicago: University of Chicago Press, 1994); Steven Shapin, *Never Pure: Historical Studies of Science as if It Was Produced by People with Bodies, Situated in Time, Space, Culture, and Society, and Struggling for Credibility and Authority* (Baltimore, MD: John Hopkins University Press, 2010).

52. Manel Brullet and Albert de Pineda, "The Design," in Manel Brullet, Albert de Pineda and Alfonso de Luna, *Parc de Recerca Biomédica de Barcelona* (Barcelona: Edicions de l'Eixample, 2007), 28.

53. Ted Hyman quoted in ZGF Architects, "J. Craig Venter Institute," p. 9 of 48, e-flyer, accessed September 20, 2015, http://issuu.com/zgfarchitectsllp/docs/j._craig_venter_institute?e=5145747/7966125.

54. Leslie, "'A Different Kind of Beauty,'" 173–221.

55. Salk Institute for Biological Studies, "The Legacy of Jonas Salk," *Inside Salk*, August 2014, 10.

56. Robert E. Kohler, *Partners in Science: Foundations and Natural Scientists, 1900–1945* (Chicago: University of Chicago Press, 1991).

57. Claudia Cohen, who died in 2007, was a former wife of Ronald O. Perelman, a billionaire New York businessman. He acquired the right to rename the building when he donated USD 20 million to his alma mater in 1995 and then exercised his naming right in 2008, following Cohen's death from ovarian cancer. When the couple divorced in 1994, Cohen was awarded USD 80 million in the divorce settlement. Cohen graduated from the University of Pennsylvania in 1972 with a bachelor's degree in communications. The name change was not well received by University of Pennsylvania academics. See Alex Williams, "At Penn, the Subject Is Gossip," *New York Times*, July 6, 2008, http://www.nytimes.com/2008/07/06/fashion/06penn.html?_r=0.

58. James Logan (Statesman), Wikipedia, https://en.wikipedia.org/wiki/James_Logan_(statesman).

59. Nicholas Basbanes, *A Gentle Madness* (New York: Henry Holt, 1995).

60. Williams, "At Penn, the Subject Is Gossip."

Part I: Boundaries

1. Bruno Latour, "Give Me a Laboratory and I Will Raise the World," in *Science Observed: Perspectives on the Social Study of Science*, ed. Karin D. Knorr-Cetina and Michael Mulkay (London: Sage, 1983), 168.

2. Disciplinarity is understood as the structural demarcation of knowledge and information (discourse) and, following Michel Foucault, the demarcations that are themselves constructed by knowledge. Foucault concentrated on the notion of "discourse" or the "discursive families" that constitute a discipline. In *The Order of Things*, Foucault discusses several naturalists, including Comte de Buffon (French eighteenth century) and Darwin (British nineteenth century), as belonging to the same "discourse." Michel Foucault, *The Order of Things: An Archaeology of the Human Sciences*, trans. Alan Sheridan (New York: Vintage, 1973), xii–xiii.

3 Islands

1. Paul A. Bove, "The End of Humanism: Michel Foucault and the Power of Disciplines," *Humanities in Society* 3 (Winter 1980): 23–40; Jan Goldstein, "Foucault among the Sociologists: The 'Disciplines' and the History of the Professions," *History and Theory* 23, no. 2 (1984): 170–192.

2. "Farm" was dropped from the Janelia Research Center's name in 2014, but "Janelia Farm" is how the research center is commonly referred to amongst scientists.

3. HHMI Howard Hughes Medical Institute, "About," accessed October 18, 2018, https://www.hhmi.org/about.

4. The reclusive billionaire Howard Hughes created HHMI as a tax shelter in the waning years of his life. Hughes gave the foundation an ambiguous mission statement: that its money must be used for "the promotion of human knowledge within the field of basic sciences," which should "probe the genesis of life itself." (HHMI Howard Hughes Medical Institute, "About," accessed October 18, 2018, https://www.hhmi.org/about.) Hughes died intestate, and after several years of legal disputes, in 1984, the organization was finally granted USD 5 billion. With that as seed money, the HHMI grew steadily over the next twenty years, and by 2011 only the Bill and Melinda Gates Foundation had a larger endowment.

5. The Janelia property was purchased by HHMI from the Dutch software maker Baan Companies in December 2000.

6. United States Department of the Interior, National Park Service, National Register of Historic Places, Inventory-Nomination Form for Janelia, 1987, https://www.dhr.virginia.gov/VLR_to_transfer/PDFNoms/053-0084_Janelia_1987_Final_Nomination.pdf.

7. Robert Pickens was a member of the North American Lily Society. Obituary, Robert S. Pickens, Former Author, Farmer in Loudoun, *Washington Post*, November 14, 1978.

8. Howard Hughes Medical Institute, *Janelia Farm: Report on Program Development* (November 2003), 4, https://www.janelia.org/sites/default/files/About%20Us/JFRC.pdf.

9. Child and infant care would be available on site to "ensure the ability to carry out high-level research while participating actively in family life." HHMI, *Janelia Farm*, 25.

10. The Salk Institute is a conspicuous absence in the reports of the HHMI, given that it too was conceived in relative isolation and is an independent not-for-profit organization outside traditional academic structures.

11. HHMI, *Janelia Farm*, 3.

12. HHMI, *Janelia Farm*, 63.

13. HHMI, *Janelia Farm*, 9.

14. HHMI, *Janelia Farm*, 39.

15. HHMI, *Janelia Farm*, 3.

16. Michel Serres, *Hermes: Literature, Science, Philosophy* (Baltimore, MD: Johns Hopkins University Press, 1982), 83. Serres was to later suggest that such pockets of order were less "islands" than oceans. See Michel Serres and Bruno Latour, *Conversations on Science, Culture, and Time*, trans. Roxanne Lapidus (Ann Arbor: University of Michigan Press, 1995), 16, 129–130; Michel Serres, "Literature and the Exact Sciences," *Substance: A Review of Theory and Literary Criticism* 18, no. 2 (1989): 3–34.

17. Serres, *Hermes*, 83.

18. Serres, *Hermes*, 82.

19. Serres and Latour, *Conversations on Science, Culture, and Time*, 129.

20. Serres and Latour, *Conversations on Science, Culture, and Time*, 129–130.

21. Bruno Latour, *Reassembling the Social: An Introduction to Actor Network Theory* (Oxford: Oxford University Press, 2005), 245.

22. H. G. Wells, *The Island of Dr Moreau*, Planet e-books, 10, accessed June 8, 2015, http://www.planetebook.com/ebooks/The-Island-of-Doctor-Moreau.pdf.

23. Wells, *The Island of Dr Moreau*, 3.

24. Charles Darwin, *On the Origin of Species by Means of Natural Selection, or the Preservation of Favoured Races in the Struggle for Life* (1859), facs. ed., intro. Ernst Mayr (Cambridge, MA: Harvard University Press, 1966), 105.

25. Bruno Latour, "Give Me a Laboratory and I Will Raise the World," in *Science Observed: Perspectives on the Social Study of Science,* ed. Karin Knorr-Cetina and Michael Mulkay (London: Sage, 1983), 141–170.

26. Peter Sloterdijk, *Spheres III–Foams, Plural Spherology* (South Pasadena, CA: Semiotext(e), 2016), 295; this is a translation of *Sphären III—Shaume* (Frankfurt: Suhrkamp, 2004) by Weiland Hoban.

27. Peter Sloterdijk, cited in Jean-Pierre Couture, "Sphere-Effects: A History of Peter Sloterdijk's Political Architectures," *World Political Science Review* 7, no. 1 (2011): 3; the reference is to Peter

Sloterdijk, *Nicht gerettet: Versuche nach Heidegger* (Frankfurt: Surkamp, 2001), 172, translation by Jean-Pierre Couture.

28. Sloterdijk, *Spheres III–Foams*, 25.

29. Sloterdijk, *Spheres III–Foams*, 309–490.

30. Sloterdijk, *Spheres III–Foams*, 296.

31. Owen Hannaway, "Laboratory Design and the Aim of Science: Andreas Libavius versus Tycho Brahe," *Isis* 77, no. 4 (1986): 609–610.

32. Steven Shapin, "'The Mind Is Its Own Place': Science and Solitude in Seventeenth-Century England," *Science in Context* 4, no. 1 (April 1991): 191.

33. Peter Sloterdijk, *Spheres II–Globes, Macrospherology* (South Pasadena, CA: Semiotext(e), 2014), 237; this is a translation of *Sphären II—Globen, Makrosphärologie* (Frankfurt: Suhrkamp, 1999) by Weiland Hoban.

34. Sloterdijk, *Spheres III–Foams*, 321–332.

35. Despite its name, Biosphere 2 had no real precursor. It was called "Biosphere 2" because Biosphere 1 was the Earth itself. Where Biosphere 1 is an island in the sky, Biosphere 2 is an island in the desert.

36. Sloterdijk, *Spheres III–Foams*, 25.

37. "HHMI Janelia Farm Research Campus, Rafael Viñoly," *arcspace*, September 18, 2006, http://www.arcspace.com/features/rafael-vinoly-/hhmi-janelia-farm-research-campus/.

38. Gerald Rubin, "Janelia Farm: An Experiment in Scientific Culture," *Cell* 125, Iss. 2, (2006): 209.

39. "Card Lab Research," HHMI Janelia Research Campus, accessed November 13, 2018, https://www.janelia.org/.

40. HHMI, *Janelia Farm*, 13.

41. Rubin, "Janelia Farm," 209.

42. Mitchell Waldrop, "Research at Janelia: Life on the Farm," *Nature* 479, no. 7373, November 16, 2011, http://www.nature.com/news/research-at-janelia-life-on-the-farm-1.9373.

43. Waldrop, "Research at Janelia."

44. Waldrop, "Research at Janelia."

45. Mac Carey, "Brave New World," *Virginia Living*, February 9, 2011, http://www.virginialiving.com/culture/brave-new-world2/.

46. Latour, "Give Me a Laboratory and I Will Raise the World," 143, 144.

47. Latour, "Give Me a Laboratory and I Will Raise the World," 150.

48. Latour, "Give me a Laboratory and I Will Raise the World," 160.

49. Rubin, "Janelia Farm," 211.

50. Rubin, "Janelia Farm," 211.

51. Gerald Ruben quoted in "John Harvard's Journal: The Janelia Experiment," *Harvard Magazine*, January–February 2007, http://harvardmagazine.com/2007/01/the-janelia-experiment.html.

52. Rubin, "Janelia Farm," 212.

4   Transparency

1. The Janelia Archives, "Artifact: Structural Glass," accessed October 24, 2018, https://www .janelia.org/archive/structural-glass.

2. The Janelia Archives, "Artifact: Structural Glass."

3. An earlier reference occurs in an 1868 address given by the Baptist minister Silas Mead, who speaks of "Adelaide, the City of Churches," *South Australian Register* (May 12, 1868), 2.

4. In the 2014 Architecture Awards of the South Australian Chapter of the Australian Institute of Architects, the SAHMRI won the Jack McConnell Award for Public Architecture, the Keith Neighbour Award for Commercial Architecture, the Derrick Kendrick Award for Sustainable Architecture, the Robert Dickson Award for Interior Architecture, and the Colorbond Award for Steel Architecture.

5. Ranking is by *Building Design*, January 2017, https://www.bdonline.co.uk.

6. Woods Bagot have a thousand staff. In 2009 the firm was named International Practice of the Year by *Architect's Journal*. The firm's origins can be traced back to 1869, when architect Edward John Woods was commissioned to improve and expand the design of St. Peter's Cathedral in Adelaide.

7. It is the first time an Australian building has won "Laboratory of the Year." In 1990 a laboratory in Perth got a "special mention."

8. Tim Studt, "More than Just a Pretty Face," *Laboratory Design*, June 10, 2016, https://www .labdesignnews.com/article/2015/06/more-just-pretty-face.

9. Steffen Lehmann, "The SAHMRI: Performance Driven," *ArchitectureAU*, January 23, 2014, http:// architectureau.com/articles/beyond-parametricism-transforming-the-city-with-sustainable-design/.

10. Carolyn Boyd, "Building Report: Building on Success," *Australian Way: Qantas*, April 1, 2014, 112.

11. Philip Stevens, "Woods Bagot Clads SAHMRI in Articulated Transparent Façade," *Design Boom*, November 29, 2013, http://www.designboom.com/architecture/woods-bagot-clads-sahmri -in-articulated-transparent-facade-11-29-2013/.

12. Woods Bagot, "News. SAHMRI: Deconstructed," accessed August 21, 2016, http://www .woodsbagot.com/news/sahmri-deconstructed.

13. John Byleveld, "Why the Fuss about SAHMRI's Pinecone?" *InDaily: Adelaide's Independent News*, September 16, 2013, http://indaily.com.au/arts-and-culture/design/2013/09/16/why-the-fuss-about-sahmris-pinecone/.

14. Belinda Willis, "SAHMRI Building a Model of Leading Architectural Design," *Advertiser*, November 29, 2013, http://www.adelaidenow.com.au/business/sahmri-building-a-model-of-leading-architectural-design/story-fni6uma6-1226770495974.

15. David Neustein, "Hi-Tech Architecture in Adelaide," *Monthly*, March 2014, https://www.themonthly.com.au/issue/2014/march/1393592400/david-neustein/hi-tech-architecture-adelaide.

16. Claire Suttles, "Adelaide's Architectural Icon," *Construction in Focus*, April 21, 2015, http://www.constructioninfocus.com.au/index.php/2015/04/21/adelaides-architectural-icon/.

17. Jamie Munro, "SAHMRI by Woods Bagot Looks like a Living Organism," *Trendhunter: Art and Design*, April 17, 2014, http://www.trendhunter.com/trends/sahmri.

18. Boyd, "Building Report," 112.

19. "A Look inside the SAHMRI building," *SA Life: Channel 7 Television*. http://www.salife7.com.au/adelaide/places/properties/a-look-inside-the-sahmri-building.

20. Willis, "SAHMRI Building a Model of Leading Architectural Design."

21. Woods Bagot, "The SAHMRI Set to Open," *ArchitectureAU*, 26 November, 2013, http://architectureau.com/articles/sahmri-set-to-open/.

22. "New Biochemistry Building Opens at the University of Oxford," *University of Oxford*, press release, December 2008, http://www.bioch.ox.ac.uk/aspsite/whatsnew/hb_oxford_biochemistry_111208.pdf.

23. "Laboratory and Logistics Building/ Mikkelsen Architects," *Archdaily*, November 12, 2018, https://www.archdaily.com/905569/laboratory-and-logistics-building-mikkelsen-architects.

24. The Sainsbury Laboratory, University of Cambridge, accessed November 13, 2012. http://www.slcu.cam.ac.uk/outreach/plantpower.html.

25. Ruth Stevens, "Elements of New Frick Lab Join to Create 'Best Infrastructure' for Chemistry," *News at Princeton*, September 2, 2010, http://www.princeton.edu/main/news/archive/S28/32/85K84/.

26. "Allen Institute, Perkins and Will," *Archdaily*, December 3, 2015, http://www.archdaily.com/778225/allen-institue-perkins-plus-will.

27. Jeremy Bentham, cited in Janet Semple, *Bentham's Prison: A Study of the Panoptican Penitentiary* (Oxford: Clarendon, 1993), 116.

28. Henri Rousseau, "Dialogues," in *The Collected Writings of Rousseau*, ed. Roger D. Masters, and Christopher Kelly, (Hanover, NH: University Press of New England, 1990).

29. Paul Scheerbart and Bruno Taut, *Glass Architecture* (New York: Praeger, 1972).

30. Walter Benjamin, "Surrealism: The Last Snapshot of the European Intelligentsia," in *One Way Street and Other Writings*, trans. E. Jephcott and K. Shorter (London: Verso, 1985), 228.

31. Walter Benjamin, *Selected Writings (Volume 2 1927–1934)*, ed. Michael W. Jennings, Howard Eiland, and Gary Smith, trans. Rodney Livingstone and others (Cambridge, MA: Belknap Press of Harvard University, 1990), 734.

32. Walter Benjamin, "Experience and Poverty" (1933), in *Selected Writings (Volume 2 1927–1934)*, 731–736.

33. Hannes Meyer quoted in Rosemarie Haag Bletter, "Mies and Dark Transparency," in *Mies in Berlin*, ed. Barry Bergdoll and Terence Riley (New York: Museum of Modern Art, 2002), 355.

34. Jacques Derrida and Maurizio Ferraris, *A Taste for the Secret*, trans. Giacomo Donis (Cambridge, UK: Polity Press, 2001), 57.

35. Nigel Whiteley, "Intensity of Scrutiny and a Good Eyeful: Architecture and Transparency," *Journal of Architectural Education* 56, no. 4 (2003): 12.

36. Janet Browne, *Charles Darwin: The Power of Place*: *Volume 2 of a Biography* (London: Jonathan Cape, 2003), 241.

37. Dustin Valen, "On the Horticultural Origins of Victorian Glasshouse Culture," *Journal of the Society of Architectural Historians* 75, no. 4, (2016), 407.

38. Roger Fulford, *Hanover to Windsor* (London: Fontana-Collins, 1966), 221.

39. Quoted in John Stamper, "London's Crystal Palace and Its Decorative Iron Construction," in *Function and Fantasy: Iron Architecture in the Long Nineteenth Century*, ed. Paul Dobraszczyk and Peter Sealy (London: Routledge, 2016).

40. George Gilbert Scott, *Remarks on Secular and Domestic Architecture, Present and Future* (London: John Murray, 1857), 108–109.

41. Dostoevsky was, evidently, terrified by the universalism he saw ahead. In his *Winter Notes on Summer Impressions* (1863) he refers to the Crystal Palace and the International Exposition it housed as a "terrible force that has united all the people here, who come from all over the world, into a single herd." Fyodor Dostoevsky, *Winter Notes on Summer Impressions*, trans. David Patterson (Evanston, IL: Northwestern University Press, 1997), 37.

42. Dostoevsky is mentioned in the novel *We* as a writer whom the servile literature of the future totalitarian state has surpassed. Yevgeny Zamyatin, *We*, trans. and foreword Gregory Zilboorg (New York: E. P. Dutton, 1959), 42.

43. Zamyatin, a Russian naval engineer and satirist, was exiled under the czarist regime in 1905 to Siberia for his Bolshevism, and again for two years in 1911. Abroad in England in 1916, Zamyatin missed the February Revolution and returned during the October Revolution. Although he had initially supported the Russian Revolution, he came to disagree more and more with its policies.

Zamyatin died in 1936 in Paris, where his friend and publisher Marc Slonim described him as "thin, almost transparent" in the weeks before succumbing to a heart ailment and poverty. Marc Slonom, Preface, Yevgeny Zamyatin, *We*, trans. and foreword Gregory Zilboorg (New York: E. P. Dutton, 1959), xxv.

44. Zamyatin, *We*, 3.

45. Zamyatin, *We*, 31–32.

46. We have used the Clarence Brown translation of this passage as it conveys a stronger sense of visual penetration; all other references are from the Zilboorg translation. Yevgeny Zamyatin, *We*, trans. and intro. Clarence Brown (London: Penguin, 1993), 64.

47. Zamyatin, *We*, 31.

48. Zamyatin, *We*, 129.

49. Zamyatin, *We*, 3.

50. Zamyatin, *We*, 44.

51. Zamyatin, *We*, 48.

52. Zamyatin, *We*, 54.

53. Zamyatin, *We*, 55.

54. Fyodor Dostoevsky, *Notes from Underground*, trans. Constance Garnett (New York: MacMillan, 1918), pt. 1, chap. 7, http://web.archive.org/web/20110116033046/http://etext.lib.virginia.edu /toc/modeng/public/DosNote.html.

55. Charles Jencks, *Modern Movements in Architecture* (London: Pelican, 1973), 200.

56. Reinholdt Martin, *The Organizational Complex: Architecture, Media and Corporate Space*, (Cambridge, MA: MIT Press, 2003), 5.

57. Martin, *The Organizational Complex*, 7.

58. Bruno Latour, "What If We Talked Politics a Little?" *Contemporary Political Theory* 2 (2003): 147.

59. Latour, "What If We Talked Politics a Little?" 143–164.

60. Bruno Latour, *On the Modern Cult of the Factish Gods* (Durham, NC: Duke University Press, 2010), 32.

61. Sammy Medina, "Architects Bring Sunshine to Nanotech Labs," *CoDesign*, August 11, 2013. http://www.fastcodesign.com/3020945/architects-give-dweeby-nanotechnology-department-a -splashy-new-outpost.

62. Medina, "Architects Bring Sunshine to Nanotech Labs."

63. Jacques Derrida, *Glas* (1974), trans. John P. Leavey and Richard Rand (Lincoln: University of Nebraska Press, 1986), 60.

64. Sydney Brenner wanted to study the processes of genetic replication and transcription, genetic recombination and mutagenesis, and the synthesis of enzymes. Brenner sought understanding of the nervous system in particular. Initially he selected for study the tiny soil-dwelling roundworm *Caenorhabditis elegans*, one of millions of species of nematode, because it is extremely simple. It contains only around a thousand cells—yet is still a real animal with nerves, muscles, intestines, and reactions. Its simplicity meant it could be cloned, sequenced, hybridized. Its entire life cycle is a mere three to seven days and 100,000 of them can live in a petri dish. It can be frozen in liquid nitrogen and stored. What's more, it's a self-fertilizing hermaphrodite yet can also reproduce by cross-fertilization, so sexual reproduction is independent of the size of the population. All of these characteristics recommended it to the scientist, and, aided with the invention of a platinum work pick, he "took to worm genetics like ducks to water" (Jonathan Hodgkin, "Early Worms," *Genetics* 121, no. 1 [1989]: 2).

65. Hodgkin, "Early Worms," 1, 2.

66. Leslie Roberts, "The Worm Project," *Science* 249, no. 4961 (June 15, 1990): 1310.

67. Andrew Brown, *In the Beginning Was the Worm: Finding the Secrets of Life in a Tiny Hermaphrodite* (London: Simon and Schuster, 2003): 9.

68. Roberts, "The Worm Project," 1311.

69. The nematode has just 302 neurons and 7,000 connections between those neurons, whereas the human brain contains around 86 billion neurons and 100 trillion synapses.

70. The genome sequence of the zebrafish was released in 2011, and, according to a paper published in *Nature* in 2013, 70 percent of protein-coding human genes are related to genes found in the zebrafish, and 84 percent of genes known to be associated with human disease have a zebrafish counterpart. Analysis of zebrafish genes and their mammalian counterparts has identified more than forty groups of genes that are syntenic (i.e., on the same chromosome) in both zebrafish and humans (Kerstin Howe et al., "The Zebrafish Reference Genome Sequence and Its Relationship to the Human Genome," *Nature* 496 [2013], 498–503). Scientists are using zebrafish to learn about the genes responsible for muscle disorders, heart defects, and nervous system diseases like multiple sclerosis (Karl J. Clark and Stephen C. Ekker, "How Zebrafish Genetics Informs Human Biology," *Nature Education* 8, no. 4 [2015]: 3).

71. Richard Mark White, Anna Sessa, Christopher Burke, Teresa Bowman, Jocelyn LeBlanc, Craig Ceol, Caitlin Bourque, Michael Dovery, Wolfram Goessling, Caroline Erter Burns and Leonard Zon, , "Transparent Adult Zebrafish as a Tool for in Vivo Transplantation Analysis," *Cell Stem Cell* 2, no.2 (2008): 187.

72. George W. Corner, *A Biographical Memoir of Warren Harmon Lewis, 1870–1964* (Washington, DC: National Academy of Sciences, 1967), 339.

73. Richard Doyle, *On Beyond Living: Rhetorical Transformations of the Life Sciences* (Stanford, CA: Stanford University Press, 1997), 20.

74. Melinda Wenner, "The Most Transparent Research," *Nature Medicine* 15 (2009): 1106.

75. Wenner, "The Most Transparent Research," 1106.

76. Zamyatin, *We*, 69.

77. Jim Endersby, *A Guinea Pig's History of Biology: The Plants and Animals Who Taught Us the Facts of Life* (London: William Heinemann, 2007), 409–410.

78. Commerical & General, *Australian Bragg Centre for Proton Therapy* homepage, accessed October 31, 2018, https://www.australianbraggcentre.com/.

5   Unbounded

1. Bruno Latour, "Give Me a Laboratory and I Will Raise the World," in *Science Observed: Perspectives on the Social Study of Science*, ed. Karin D. Knorr-Cetina and Michael Mulkay (London: Sage, 1983), 168.

2. Latour, "Give Me a Laboratory and I Will Raise the World," 168.

3. Latour, "Give Me a Laboratory and I Will Raise the World," 166–167.

4. Stephen Zacks, "The DNA of Science Labs," *Metropolis Magazine*, February 2007, http://www.metropolismag.com/February-2007/The-DNA-of-Science-Labs/.

5. Leonardo da Vinci, quoted in Avrum Stroll, *Surfaces* (Minneapolis: University of Minnesota Press, 1988), 75–76.

6. Stroll, *Surfaces*, 39–40.

7. Latour, "Give Me a Laboratory and I Will Raise the World," 168.

8. Paolo Fumagalli, "The Chipperfield Building in Its Context—Interpretation after Viewing," in *Novartis Campus—Fabrikstraße 22: David Chipperfield*, ed. Ulrike Jehle-Schulte Strathaus (Basel: Christoph Merian Verlag, 2011), 17.

9. Rowan Moore, "David Chipperfield: A Master of Permanence Comes Home," *Observer*, February 6, 2011, http://www.guardian.co.uk/artanddesign/2011/feb/06/david-chipperfield-turner-contemporary-hepworth-wakefield.

10. Novartis was seventh in the previous report from 2012. The report is funded by the Bill and Melinda Gates Foundation, the UK Department for International Development, and the Dutch Ministry of Foreign Affairs. Access to Medicine Foundation, *Access to Medicine Index 2014* (Haarlem, The Netherlands: Access to Medicine Foundation, 2014), 11, http://apps.who.int/medicinedocs/documents/s19987en/s19987en.pdf.

11. Tufts Center for the Study of Drug Development, "Cost to Develop and Win Marketing Approval for a New Drug Is $2.6 Billion," November 18, 2014, http://csdd.tufts.edu/news/complete_story/pr_tufts_csdd_2014_cost_study.

12. Denise Roland, "Novartis Pharmaceuticals Head to Depart amid Restructure," *Wall Street Journal*, May 17, 2016, accessed November 24, 2016, http://www.wsj.com/articles/novartis-to-split-pharmaceuticals-division-into-two-1463506092.

13. Associated Press, "Novartis 1Q Profit Jumps 12 Percent as Heart Drug Sales Soar," *Seattle Times*, April 19, 2018, https://www.seattletimes.com/business/novartis-1q-profit-jumps-12-percent-as -heart-drug-sales-soar/.

14. New pharmaceutical companies, in Basel especially, have followed the example set by Novartis. Actelion, for example, commissioned Herzog and de Meuron for its headquarters (2010) and laboratories (2012). Since Vasella's departure, Novartis has expanded its commissioning practices to include upcoming as well as established architects. For their new USD 600 million campus on 3.76 acres in the heart of Kendall Square in Cambridge, Novartis appointed three architectural practices that are led and owned by women of Asian ethnicity. An eight-story laboratory tower with vivarium was designed by Toshiko Mori, a professor at Harvard's Graduate School of Design. Maya Lin, architect of the Vietnam War Memorial, designed a four-story office building with auditorium, cafeteria, and retail space. Renovations for a third building were undertaken by the Boston and Seoul–based Single Speed Design, the emerging practice of Jinhee Park and John Hong. While Mori has previously undertaken large institutional and corporate projects, none of these three has experience in the design of laboratories necessitating biological containment certification. Each is known for her artistic sensibility and academic underpinnings, and their appointment is clearly intended to signal progressive values, including a commitment to gender and racial equity.

15. Vittorio Magnano Lampugnani is professor for the History of Urban Design at the Swiss Federal Institute of Technology in Zurich and has his own architectural practice in Milan (Studio di Architettura) as well as another one in Zurich (Baukontor Architekten) with two associate partners. He continues to be a member of the Novartis Campus Steering Committee, Basel, as well as a member of the Novartis Campus Steering Committees for East Hanover, New Jersey, Cambridge, Massachusetts, and Shanghai.

16. Lampugnani delivered the lecture at the ICAM (International Confederation of Architectural Museums) Conference held at the Center for Architecture in Vienna on September 24, 2002. A revised version was published 2006 as Vittorio Magnano Lampugnani, "Insight versus Entertainment: Untimely Meditations on the Architecture of Twentieth-Century Art Museums," in *A Companion to Museum Studies*, ed. Sharon MacDonald (Malden, MA: Blackwell, 2006), 245–262.

17. Novartis homepage, accessed September 30, 2011, http://campus.novartis.com//.

18. Nicolai Ouroussoff, "Many Hands, One Vision," *New York Times*, December 23, 2009, http:// www.nytimes.com/2009/12/27/arts/design/27novartis.html?_r=0.

19. Vasella quoted in Ouroussoff, "Many Hands, One Vision."

20. The tours take place on select Saturdays. A private guided tour with an English-speaking guide is available for CHF 400 (around USD 400) and can cater for up to twenty people. The public guided tour costs CHF 22 for adults and is in German only.

21. Vasella quoted in Ouroussoff, "Many Hands, One Vision."

22. Ouroussoff, "Many Hands, One Vision."

23. Ouroussoff, "Many Hands, One Vision."

24. Fumagalli, "The Chipperfield Building in Its Context," 17.

25. The previous laboratories are 16 Fabrikstraße by Adolf Krischanitz (2004–2008), 14 Fabrikstraße by Rafael Moneo (2005–2009), and 10 Fabrikstraße by Yoshio Taniguchi (2005–2010).

26. Ruth Reed, quoted in "RIBA Gold Medal-2011: David Chipperfield," *e-architect*, February 9, 2011, http://www.e-architect.co.uk/awards/riba_gold_medal_2011.htm.

27. Burkhardt and Partner were also architects in association for the Novartis laboratory WSJ-352 by Tadao Ando on the Basel campus.

28. Bernhard Fibicher, "Serge Spitzer's Molecular Topography," in *Novartis Campus–Fabrikstraße 22: David Chipperfield*, ed. Ulrike Jehle-Schulte Strathaus (Basel: Christoph Merian Verlag, 2011), 65.

29. David Chipperfield Architects, "Inventory," in *Novartis Campus–Fabrikstraße 22: David Chipperfield*, ed. Ulrike Jehle-Schulte Strathaus (Basel: Christoph Merian Verlag, 2011), 77.

30. Ulrike Jehle-Schulte Strathaus, "Introduction," in *Novartis Campus–Fabrikstraße 22: David Chipperfield*, ed. Ulrike Jehle-Schulte Strathaus (Basel: Christoph Merian Verlag, 2011), 9.

31. Mark Fishman, "The Future, Today," in *Novartis Campus-Fabrikstraße 22: David Chipperfield*, ed. Ulrike Jehle-Schulte Strathaus (Basel: Christoph Merian Verlag, 2011), 7.

32. The history and consequences of this assumption in laboratory design is discussed in Sandra Kaji-O'Grady, "The Spaces of Experimental Science," *Journal Spéciale'Z* 4 (2012): 88–99.

33. Peter Sloterdijk, *Spheres III–Foams, Plural Spherology* (South Pasadena: Semiotext(e), 2016), 295; this is a translation of *Sphären III—Shaume* (Frankfurt: Suhrkamp, 2004), by Weiland Hoban.

34. Sloterdijk, *Spheres III-Foams*, 18–19.

35. Toshiko Mori Architect, "Novartis Institutes for BioMedical Research Cambridge, MA," accessed October 2, 2016, http://www.tmarch.com/2677.

36. David Chipperfield Architects, "Inventory," 77.

37. Bill Hillier and Alan Penn, "Visible Colleges: Structure and Randomness in the Place of Discovery," *Science in Context* 4, no. 1 (1991): 45.

38. "Interview with Ross Lovegrove," *Designboom*, October 31, 2006, http://www.designboom.com/interviews/designboom-interview-ross-lovegrove/.

39. David Chipperfield Architects, "Inventory," 77.

40. Gilbert Simondon, *On the Mode of Existence of Technical Objects* (Paris: Aubier, Editions Montaigne, 1958), trans. Ninian Mellamphy (University of Western Ontario, 1980), 47–48. https://english.duke.edu/uploads/assets/Simondon_MEOT_part_1.pdf.

41. The term was coined by William Wulf in 1989 while he worked at the US National Science Foundation. Wulf quotes himself in William Wulf, "The Collaboratory Opportunity," *Science* 261 (August 1993): 854.

42. Richard Zare, "Knowledge and Distributed Intelligence," *Science* 275, no. 5303 (1997): 1047.

43. Eric Winsberg, "Simulated Experiments: Methodology for a Virtual World," *Philosophy of Science* 70, no. 1 (2003): 109.

44. Przemyslaw Prusinkiewicz, "Art and Science of Life: Designing and Growing Virtual Plants with L-Systems," in *Nursery Crops: Development, Evaluation, Production and Use: Proceedings of the XXVI International Horticultural Congress*, eds. C. Davidson and T. Fernandez, *Acta Horticulturae (ISHS)* 630, (2004), 15–28, accessed February 14, 2013, http://www.actahort.org/books/630/630_1 .htm.

45. K. Takashi, K. Yugi, K. Hashimoto, Y. Yamada, C. J. F. Pickett, and M. Tomita, "Computational Challenges in Cell Simulation: A Software Engineering Approach," *Intelligent Systems, IEEE* 17, no. 5, (2002): 64–71.

46. Masura Tomita, "Whole-Cell Simulation: A Grand Challenge of the 21st Century," *Trends in Biotechnology* 19, no. 6 (2001): 205–210.

47. G. Desmuelles, G. Querrec, P. Redou, S. Kordelo, L. Misery, V. Rodin, and J. Tisseau, "The Virtual Reality Applied to Biology Understanding: The in Virtuo Experimentation," *Expert Systems with Applications* 30, no. 1 (2006): 82–92.

48. Karin Knorr Cetina, "The Couch, the Cathedral, and the Laboratory," in *Science as Practice and Culture*, ed. Andrew Pickering (Chicago: University of Chicago Press, 1992), 113–138.

49. Knorr Cetina, "The Couch," 117.

50. Joan Fujimura, "Technobiological Imaginaries: How Do Systems Biologists Know Nature?," in *Knowing Nature: Conversations at the Intersection of Political Ecology and Science Studies*, ed. M. J. Goldman, P. Nadasdy, and M. D. Turner (Durham, NC: Duke University Press, 2011), 69.

51. Sarah Neville and Ralph Atkins, "Novartis's New Chief Sets Sights on 'Productivity Revolution,'" *Financial Times*, September 26, 2017, https://www.ft.com/content/5ab8ba6e-9c7a -11e7–9a86–4d5a475ba4c5.

52. Neville and Atkins, "Novartis's New Chief Sets Sights on 'Productivity Revolution.'"

53. Vikas Dandekar and C R Sukumar, "There's an Explosion of Data and Digital Opportunities in Indian Healthcare: Vasant Narasimhan, CEO, Novartis," *Economic Times*, August 21, 2018, https://economictimes.indiatimes.com/industry/healthcare/biotech/pharmaceuticals/theres-an -explosion-of-data-digital-opportunities-in-indian-healthcare-vasant-narasimhan-ceo-novartis /articleshow/65481279.cms.

54. Peter Sloterdijk, "The Crystal Palace" (2005), trans. Michael Darroch, *Public* 37 (2009): 15. https://pi.library.yorku.ca/ojs/index.php/public/article/viewFile/30252/27786.

55. Peter Sloterdijk, "The Crystal Palace," 15.

6 Enunciation

1. "Blizard Building," arcspace.com, November 21, 2005, http://www.arcspace.com/features/alsop-architects/blizard-building/.

2. Gilles Deleuze, "Postscript on the Societies of Control," *October* 59 (Winter, 1992): 3–7. This essay first appeared in *L'Autre journal*, no. 1 (May 1990).

3. Gilles Deleuze, *The Logic of Sense*, trans. Mark Lester and Charles Stivale (London: The Athlone Press, 1990), 125.

4. Deleuze, "Postscript on the Societies of Control," 4.

5. Chris L. Smith and Sandra Kaji O'Grady, "Exaptive Translations between Biology and Architecture," *Architectural Research Quarterly (ARQ)* 18, no. 2 (2014): 155–166.

6. John Tyler Bonner, "Analogies in Biology," *Synthese* 15, no. 2 (1963): 275.

7. John Tyler Bonner, *Randomness in Evolution* (Princeton, NJ: Princeton University Press, 2013).

8. Stephen Jay Gould, "D'Arcy Thompson and the Science of Form," *New Literary History* vol. 2, no. 2 (1971): 233–234.

9. The "unit" of evolution for morphologists was most definitely the organism. For a critique of the changing notion of "organism" within biological thought, see Keith R. Benson, "Biology's 'Phoenix': Historical Perspectives on the Importance of the Organism," *American Zoology* 29, no. 3 (1989): 1067–1074.

10. D'Arcy Wentworth Thompson, *On Growth and Form* (Cambridge, UK: Cambridge University Press, 2004), 3.

11. Philip Steadman, *The Evolution of Designs: Biological Analogy in Architecture and the Applied Arts* (Abingdon, UK: Routledge, 2008), 12.

12. For a fuller discussion of Wentworth Thompson's influence, see Philip Beesley and Sarah Bonnemaison, eds., *On Growth and Form: Organic Architecture and Beyond* (Toronto: Riverside Architectural Press, 2008).

13. Nicholas Negroponte, *Architecture Machine* (Cambridge, MA: MIT Press, 1970); Greg Lynn, *Animate Form*, (Princeton, NJ: Princeton University Press, 1998).

14. Bonner, "Analogies in Biology," 276.

15. Steadman, *Evolution of Designs*, 95. Steadman quotes from Richard Owen, *The Archetype and Homologies of the Vertebrate Skeleton* (London: John Van Voorst, 1848).

16. Kaja Silverman, *The Subject of Semiotics* (New York: Oxford University Press, 1983), 10.

17. Ferdinand de Saussure, *Course in General Linguistics*, trans. Wade Baskin (London: Fontana/Collins, 1974), 122.

18. Alan Bullock and Stephen Trombley ed., *The Norton Dictionary of Modern Thought* (New York: W.W. Norton, 1999), 27.

19. Bullock and Trombley, *The Norton Dictionary of Modern Thought*, 27.

20. Fredric Jameson, *The Political Unconscious: Narrative as a Socially Symbolic Act* (Ithaca, NY: Cornell University Press, 1981), 13.

21. Steadman, *Evolution of Designs*, vi.

22. Steadman, *Evolution of Designs*, xv.

23. Steadman, *Evolution of Designs*, xvi.

24. Steadman, *Evolution of Designs*, xvi.

25. All Design, "Centre for the Cell: 'Neuron Pod,' The Blizard Institute Queen University of London," Planning Report Incorporating Design and Access Statement, April 2012, section 3.

26. Blizard Institute Homepage, "Core Facilities and Laboratory Management," accessed November 11, 2016, http://www.blizard.qmul.ac.uk/research/core-facilities/11-blizard-institute/blizard.html.

27. Cited in All Design, "Neuron Pod," Planning Application (April 2012).

28. de Saussure, *Course in General Linguistics*, 128.

29. Tom Porter, *Will Alsop—The Noise* (London: Routledge, 2011), 113.

30. Porter, *Will Alsop*, 113.

31. Legend has it that Alsop made his commitment to blobs known at a Royal Academy Forum. Will Alsop RA (1947–2018), Royal Academy homepage, accessed November 1, 2018, https://www.royalacademy.org.uk/art-artists/name/will-alsop-ra.

32. John Tyler Bonner, *Why Size Matters: From Bacteria to Blue Whales* (Princeton, NJ: Princeton University Press, 2006).

33. Land Design Studio, accessed November 11, 2016, http://www.landdesignstudio.co.uk/.

34. Land Design Studio, Company Profile, accessed November 11, 2016, http://www.landdesignstudio.co.uk/new_site/files/LDS_Pack_1509.pdf.

35. Richard Doyle, *On Beyond Living: Rhetorical Transformations of the Life Sciences* (Stanford, CA: Stanford University Press, 1997), 59.

36. Bonner, "Analogies in Biology," 279.

37. Bonner, "Analogies in Biology," 279.

7   Excess

1. Philip Steadman, *The Evolution of Designs: Biological Analogy in Architecture and the Applied Arts* (Cambridge, UK: Cambridge University Press, 1979).

2. Peter Collins, *Changing Ideals in Modern Architecture, 1750–1950* (Montreal: McGill-Queens University Press, 1998), 149.

3. Haeckel's drawings were originally published in sets of ten between 1899 and 1904 and collectively in two volumes in 1904, as *Kunstformen der Natur*. Over the course of his career, over 1,000 engravings were produced based on Haeckel's sketches and watercolors; many of the best of these were translated from sketch to print by lithographer Adolf Giltsch.

4. Binet's 1899 letter to Haeckel, cited in Philip Ball, *Shapes: Nature's Patterns, A Tapestry in Three Parts* (Oxford: Oxford University Press, 2009), 42. On Haeckel's apparent misrepresentation of scientific data, see Nick Hopwood's excellent analysis in "Pictures of Evolution and Charges of Fraud: Ernst Haeckel's Embryological Illustrations," *Isis* 97, no. 2 (2006): 260–301.

5. Peter J. Bowler, *The Eclipse of Darwinism: Anti-Darwinian Evolution Theories in the Decades around 1900* (Baltimore, MD: John Hopkins University Press, 1983).

6. Geoffrey Scott, *The Architecture of Humanism: A Study in the History of Taste* (London: Methuen, 1914), 165.

7. Catherine Ingraham, *Architecture, Animal, Human* (London: Routledge, 2006), 15.

8. Ingraham, *Architecture, Animal, Human*, 15.

9. "Directive 2010/63/EU of the European Parliament and of the Council of 22 September 2010 on the Protection of Animals Used for Scientific Purposes," *Official Journal for the European Union: L276*, September 20, 2010, 36.

10. "Directive 2010/63/EU," 34.

11. "Directive 2010/63/EU," 40.

12. Juan Irigaray Huarte, quoted in Lydia Parafianowicz, "Biomedical Research Centre," *Frame Magazine*, May 13, 2012,http://www.frameweb.com/news/biomedical-research-centre.

13. Georges Bataille, *The Accursed Share, Volume 1,* trans. Robert Hurley (New York: Zone Books, 1989), 10.

14. Bataille, *The Accursed Share*, 22.

15. Bataille, *The Accursed Share*, 9.

16. Bataille, *The Accursed Share*, 23.

17. Bataille, *The Accursed Share*, 24.

18. Bataille, *The Accursed Share*, 24.

19. Plants also get a brief mention along with micro-organisms and duckweed. Bataille, *The Accursed Share*, 28, 32–33.

20. Claude Lévi-Strauss, *Totemism*, trans. Robert Needham (Boston: Beacon Press, 1963), 89. In the original text Levi Strauss writes, "Les espèces sont choisies non commes bonnes à manger,

mais comme bonnes à penser" (Claude Lévi-Strauss, *Totémise* [Paris: Universitaires de France, 1962], 128).

21. Georges-Louis Leclerc, Comte de Buffon, *Histiore naturelle, Quadrupédes*, vol. 3 (Paris: A la Librairie Stéréotype de P. Didot, 1799), "Le Tigre," plate 8, 209; image by Pauquet.

22. John E. Grant, "The Art and the Argument of 'The Tyger,'" in *Discussions of William Blake*, ed. John E. Grant (Boston: Heath, 1961): 76–81.

23. Bataille, *The Accursed Share*, 33.

24. Bataille, *The Accursed Share*, 34.

25. Bataille, *The Accursed Share*, 29.

26. Bataille, *The Accursed Share*, 30.

27. Bataille, *The Accursed Share*, 35.

28. Bataille, *The Accursed Share*, 133.

29. NavarraBioMed, "Strategic Health Research Plan for Navarre," accessed June 30, 2015, http://www.navarrabiomed.es/en/about-us/strategic-plan.

30. NavarraBioMed, "Strategic Plan."

31. The government and the parliament of the Autonomous Region of Navarre approved in April 2008 a EUR 609 million investment program, which complements the 2007–2013 Operational Plan Navarra, which was approved by the European Commission in November 2007 and co-financed by the European Regional Development Fund. European Investment Bank, "Education and Health in Navarre Get a Boost," accessed June 29, 2015, http://www.eib.org/infocentre/stories/all/2015-april-01/education-and-health-in-navarre-get-a-boost.htm. Refer also to NavarraBioMed "Strategic Plan."

32. European Investment Bank, "Education and Health in Navarre Get a Boost."

33. The project manager was Daniel Galar Irurre and the project team included Borja Benavent, David Eguinoa, Oscar Martínez, Juan Carlos de la Iglesia, Ángel Álvarez, and Isabel Franco.

34. "CIB/Vaíllo & Irigaray," *ArchDaily*, May 1, 2012, https://www.archdaily.com/229821/cib-vaillo-irigaray-galar.

35. In 2008 only 28.8 percent of the population of Navarre (aged sixteen and over) had completed higher education, and in 2009 the unemployment rate in the Navarra region was 16 percent. See Ministry of Health, Social Policy and Equality, *2011 Annual Report of the National Health System of Spain*, 37, accessed June 20, 2015, http://www.msssi.gob.es/organizacion/sns/planCalidadSNS/pdf/equidad/informeAnual2009/AnnualReportNHS2009English.pdf.

36. NavarraBioMed, "Strategic Plan."

37. NavarraBioMed, "Patrocinadores y colaboradores", accessed June 20, 2015, http://www.navarrabiomed.es/es/presentacion/patrocinadores-y-colaboradores.

38. Krishna Ramanujan, "Clean, White, Open Spaces and Lots of Light: Weill Hall Opens for Business," *Cornell Chronicle*, June 6, 2008, http://news.cornell.edu/stories/2008/06/clean-white-open-spaces-and-lots-light-weill-hall-opens.

39. Walter Burkert. *Greek Religion*, trans. John Raffan, (Cambridge, MA: Harvard University Press, 1985), 82.

40. Bataille, *The Accursed Share*, 162.

41. Jed Mayer, "Ways of Reading Animals in Victorian Literature, Culture and Science," *Literature Compass* 7, no. 5 (2010): 347–357.

42. Spain's 2014 Animal Research Statistics from Ministerio de Agricultura, Alimentacion y Medio Ambiente, "Informe Sobre Usos De Animales En Experimentación Y Otros Fines Científicos, Incluyendo La Docencia En 2014," http://www.magrama.gob.es/es/ganaderia/temas/produccion-y-mercados-ganaderos/informesobreusosdeanimalesen2014_tcm7–402651.pdf.

43. Article 54.3, "Directive 2010/63/EU," 50.

44. See for example the website of Laboratory Animals Limited, http://www.lal.org.uk/pdffiles/lab1566.pdf. The United States Department of Agriculture also has a comprehensive literature resource for research animal welfare: http://awic.nal.usda.gov/nal_display/index.php?info_center=3&tax_level=1&tax_subject=169.

45. Julie Urbanik, "Locating the Transgenic Landscape: Animal Biotechnology and Politics of Place in Massachusetts," *Geoforum* 38, no. 6 (2007): 1214.

46. Lynda Birke, Arnold Arluke, and Mike Michael, *The Sacrifice: How Scientific Experiments Transform Animals and People* (West Lafayette, IN: Purdue University Press, 2007), 60.

47. Bataille, *The Accursed Share*, 46.

48. "Overall, 808,827 animal procedures were conducted in 2014, a 12% fall from 2013, but also the first statistical release under the new EU guidelines." And "The new EU guidelines also require retrospective reporting of animal suffering in experiments. Of the 808,827 procedures, 53% were subthreshold or mild, 27% were mild, 8% were severe, and 12% non-recovery (where the animal is fully anaesthetised before surgery and then never woken up)" Spain 2014, Spain's 2014 Animal Research Statistics.

49. Denis Hollier, *Against Architecture: The Writings of Georges Bataille*, trans. Betsy Wing (Cambridge, MA: MIT Press, 1989), xiv.

50. Ingraham, *Architecture, Animal, Human*, 226.

51. Birke, Arluke and Michael, *The Sacrifice*, 173.

52. Ingraham, *Architecture, Animal, Human*, 226.

53. Vaillo and Irigaray, "Biomedical Research Centre," accessed July 2, 2015, http://www.vailloirigaray.com/portfolio/biomedical-research-centre/.

54. Vaillo and Irigaray, "Concepts," accessed July 2, 2015, http://www.vailloirigaray.com/concepts/.

55. "CIB/Vaillo and Irigaray."

56. "Forschungszentrum für Biomedizin CIB in Pamplona" [Biomedical Research Center CIB in Pamplona], *Baunetz Wissen*, accessed July 2, 2015, http://www.baunetzwissen.de/objektartikel /Fassade-Forschungszentrum-fuer-Biomedizin-in-Pamplona_3213863.html.

57. Ala' Abuhasan, "Centro de Investigaciones Biológicas | Vaillo + Irigaray," *Arch20*, accessed July 2, 2015, http://www.arch2o.com/centro-de-investigaciones-biol%C3%B3gicas-vaillo-irigaray/.

58. Antoine Picon, *Ornament: The Politics of Architecture and Subjectivity* (London: John Wiley and Sons, 2013).

59. Jonathan Massey, "Ornament and Decoration," in *The Handbook of Interior Architecture and Design*, Part 3.1 *Atmospheric Conditions of the Interior*, ed. Graeme Brooker and Lois Weinthal (London: Boomsbury, 2013), 497.

60. See Jacques Soulillou, *Le Décoratif* (Paris: Klincksieck, 1990) and *Le Livre de l'ornement et de la guerre* (Marseille: Parenthèses, 2003). Soulillou, with N. Neumann, translated Semper into French; see Gottfried Semper, *Du Style et de l'architecture: Écrits, 1834–1869* (Marseille: Parenthèses, 2007).

8   Deep Time

1. Ledcor Group, "Project Profile: UBC Pharmaceutical Science Building," accessed November 27, 2016, http://www.ledcor.com/our-projects/environmental/sustainable-building/ubc-pharmaceutical -sciences-building.

2. Mark Hume, "'Cubist Tree' Blooms at New UBC Sciences Building," *Globe and Mail*, May 28, 2013, accessed June 30, 2016, http://www.theglobeandmail.com/news/british-columbia/cubist -tree-blooms-at-new-ubc-sciences-building/article12218509/.

3. Adele Weder, "Strong Medicine," *German-Architects Review*, September, 3, 2013, http://www.german -architects.com/en/projects/42660_Pharmaceutical_Building_on_the_University_of_British _Columbia_Campus.

4. Saucier quoted in Weder, "Strong Medicine." Plant-derived natural products in drug discovery have diminished in significance since the advent of synthetic drugs (e.g., Aspirin in 1897), combinatorial chemistry, and computational (in silico) drug design, but plants continue to serve as the basis for many pharmaceuticals today.

5. Gilles Saucier quoted in James Gauer, "Medicine Chest: UBC Faculty of Pharmaceutical Sciences," *Architectural Record*, May 16, 2013, http://www.architecturalrecord.com/articles/7936-ubc -faculty-of-pharmaceutical-sciences.

6. News Desk, "Building on Nature," *Montreal Gazette*, November 26, 2013, http://montrealgazette .com/life/urban-expressions/building-on-nature.

7. Juhani Pallasmaa, *The Eyes of the Skin: Architecture and the Senses*, preface by Steven Holl (Chichester, UK: Wiley Academy, 2005), 41.

8. David Gissen elaborates on architecture's relationship to the unmanageable elements of nature in *Subnature: Architecture's Other Environments* (Princeton, NJ: Princeton Architectural Press, 2009).

9. For contemporary architects committed to "form-finding" such as Michael Hensel and Michael Weinstock, nature's forms follow given principles that have evolved in the process of adaptation toward optimization. They are always adaptive, use a minimum of material and energy with maximum effect, and as such are explicable. See Joseph Lim Ee Man, *Bio-Structural Analogues in Architecture* (Amsterdam: BIS Publishers, 2009); Michael Weinstock, *The Architecture of Emergence: The Evolution of Form in Nature and Civilisation* (London: Wiley, 2010); Michael Hensel, *Emergent Technologies and Design: Towards a Biological Paradigm for Architecture* (Abingdon, UK: Routledge, 2010); Blaine Brownell, *Hypernatural: Architecture's New Relationship with Nature* (Princeton NJ: Princeton Architectural Press, 2015).

10. Many recent buildings are claimed to be based on trees and sport tree-like structures, among them, Jürgen Mayer's Metropol Parasol project for the Plaza de la Encarnacion in Seville, and Gallery Tote by Serie Architects in Mumbai, India. In Japan we would include Kengo Kuma's Yusuhara Marche (2012) and NAP Co.'s Forest Chapel (2013) in Sayama; and Junya Ishigami's Kanagawa Institute of Technology (2010) with its "forest" of slender metal posts arranged in clusters and clearings.

11. Sarah Goldhagen, "Seeing the Building for the Trees," *New York Times*, January 7, 2012, http://www.nytimes.com/2012/01/08/opinion/sunday/seeing-the-building-for-the-trees.html?_r=0.

12. Goldhagen, "Seeing the Building for the Trees."

13. INRA Research Laboratories/Tectoniques Architects, *ArchDaily*, April 3, 2013, http://www.archdaily.com/354179/inra-research-laboratories-tectoniques-architects.

14. Editorial Desk, "Durbach Block Jaggers and BVN's New Education Building for UTS Unveiled," *ArchitectureAU*, April 27, 2015, http://architectureau.com/articles/durbach-block-jaggers-and-bvns-new-education-building-for-uts-unveiled/.

15. DZNE German Center for Neurodegenerative Diseases/Wulf Architekten, *ArchDaily*, September 21, 2017, https://www.archdaily.com/880109/dzne-german-center-for-neurodegenerative-diseases-wulf-architekten.

16. Saucier and Perrotte, Internet homepage, accessed September 20, 2016, http://saucierperrotte.com/en/projects/pavillon-des-sciences-anne-marie-edward-du-college-john-abbott-college/.

17. Saucier and Perrotte, Internet homepage.

18. Saucier and Perrotte, Internet homepage.

19. News Desk, "Building on Nature."

20. News Desk, "Building on Nature."

21. Saucier quoted in Weder, "Strong Medicine."

22. Gauer, "Medicine Chest." The analogy of the tree was eagerly taken up by the clients, too, who describe the building they now inhabit as "mirroring the structure of a forest" and the building's staircases as connecting "the floors like organic branches" (Faculty of Pharmaceutical Sciences Internet homepage, "About Our Building," accessed June 30, 2016, http://pharmsci.ubc.ca /facilities/about-our-building).

23. Eva Bjerring, "UBC Faculty of Pharmaceutical Sciences/CDRD," *arcspace.com*, May 30, 2013, http://www.arcspace.com/features/saucier--perrotte/ubc-faculty-of-pharmaceutical-sciences--cdrd/.

24. Hume, "'Cubist Tree' Blooms."

25. It is an eight-story elliptical building by SIAT Architekten (International Neuroscience Institute homepage, accessed April 2, 2014, http://www.ini-hannover.de/en/architecture.html).

26. In *Learning from Las Vegas*, Robert Venturi, Steven Izenour, and Denise Scott Brown propose that buildings fall into two camps: those that look like what they are, and those that don't. The first type they call the "Duck" after a building on Long Island shaped like a duck from which its owner, a farmer, sold eggs. Their point is that some buildings combine function and symbolism seamlessly. The second they call the Decorated Shed, defined as a generic building to which ornamentation, symbols, and signs are applied. It was this approach that they preferred in their own design practice, notoriously adding an oversize, decorative television antennae to Guild House (1963) to symbolize how the elderly occupants spent their days. The dichotomy between the Duck and the Decorated Shed does not capture the full nuance of the authors' arguments, but its repercussions are still felt today. Robert Venturi, Steven Izenour and Denise Scott Brown, *Learning from Las Vegas: The Forgotten Symbolism of Architectural Form*, rev. ed. (Cambridge MA: MIT Press, 1977).

27. Charles Jencks, an enthusiastic advocate for the overtly symbolic architecture of the postmodern period, has maintained that the key to avoiding the literalness of the Duck or the more superficial expressions of the Shed is for architects to embed ambiguous and complex metaphorical allusions. What he refers to as "double-coding" arises in tandem with the diverse cultural references brought to the viewing experience by the audience. Jencks has no time for architect Farshid Moussavi's argument, made in *The Function of Style* (2014), that the absence of a shared understanding precludes the ability of built forms to convey meaning through signification. Charles Jencks, "The Iconic Building Is Here to Stay," *Hunch: The Berlage Institute Report* 11 (Winter 2006/7): 60.

28. Gilles Saucier quoted in Gauer, "Medicine Chest."

29. UBC Faculty of Pharmaceutical Sciences / Saucier + Perrotte Architects, *ArchDaily*, October 19, 2012, accessed June 20, 2016, http://www.archdaily.com/283728/ubc-faculty-of-pharmaceutical -sciences-saucier-perrotte-architectes.

30. UBC Faculty of Pharmaceutical Sciences / Saucier + Perrotte Architects.

31. Weder, "Strong Medicine."

32. Goldhagen, "Seeing the Building for the Trees."

33. Gottfried Semper, "Development of Architectural Style," *Inland Architect and News Record* 14 (1889): 77.

34. Walter Alvarez, *T-Rex and the Crater of Doom* (Princeton, NJ: Princeton University Press, 1997), 17.

35. Stephen Jay Gould, *The Lying Stones of Marrakech: Penultimate Reflections in Natural History* (New York: Harmony Books, 2000), 65.

36. Gould, *The Lying Stones of Marrakech*, 45.

37. Gould, *The Lying Stones of Marrakech*, 9–26.

38. Georges Cuvier, "Preliminary Discourse on the Revolutions of the Globe (1812)," in Martin J. S. Rudwick, *Georges Cuvier, Fossil Bones, and Geological Catastrophes* (Chicago: University of Chicago Press, 1997), 205.

39. See Charles Darwin, chapter 9, "On the Imperfection of the Geological Record," and chapter 10, "On the Geological Succession of Organic Beings," *On The Origin of Species by Means of Natural Selection, or the Preservation of Favoured Races in the Struggle for Life* (1859), facs. ed., intro. Ernst Mayr (Cambridge, MA: Harvard University Press, 1966).

40. Charles Darwin, *On The Origin of Species,* 292.

41. Michel Foucault, *The Order of Things: An Archaeology of the Human Sciences* (New York: Random House, 1970), 155.

42. Foucault, *The Order of Things*, 156–157.

43. Foucault, *The Order of Things*, 161.

44. Michel Foucault, *Speech Begins after Death*, translated by Robert Bononno (Minneapolis: Minnesota University Press, 2013).

45. The overlay of a new biographical history once a fossil enters the world of paleontology is especially contingent on ownership, a source of ethical conflict in paleontology. The US Society of Vertebrate Paleontology believes that scientifically important fossils should be held in perpetuity in the public trust, in museums or research institutions, where they can benefit the whole of the scientific community. In most nations, however, fossils collected legally by amateurs cannot be confiscated by museum authorities, nor can they be exported without a permit as they are considered natural heritage. A dearth of professional palaeontologists with the funds to source fossils has meant that amateurs and private collections are an essential part of the discipline and in the trade of artefacts and knowledge about them.

46. Bruno Latour, "A Textbook Case Revisited—Knowledge as a Mode of Existence," in *The Handbook of Science and Technology Studies*, 3rd ed., ed. E. Hackett, O. Amsterdamska, M. Lynch, and J. Wacjman (Cambridge, Mass, MIT Press, 2007), 83–112.

47. Bruno Latour, "A Textbook Case Revisited," 84.

48. Bruno Latour, *Pandora's Hope: Essays on the Reality of Science Studies* (Cambridge, MA: Harvard University Press, 1999), 298.

49. Latour, "A Textbook Case Revisited," 105.

50. Latour, "A Textbook Case Revisited," 105.

51. John Dewey, *Experience and Nature* (Chicago: Open Court, 1925), 381–382.

52. Roger Caillois, *The Writing of Stones* (1970), intro. Marguerite Yourcenar, trans. Barbara Bray (Charlottesville: University Press of Virginia, 1985), 34.

53. Roger Caillois, *The Writing of Stones*, 6.

54. Roger Caillois, *The Writing of Stones*, 26.

55. Roger Caillois, *La Pieuvre: Essai sur la Logique de l'Imaginaire* (Paris: La Table Ronde, 1973), 229, quoted in Marina Warner, "The Writing of Stones: Roger Caillois's Imaginary Logic," *Cabinet* 29, (2008): 34–41, http://cabinetmagazine.org/issues/29/warner.php.

56. In *Stones of Venice* (1853), Ruskin intensely metaphorizes the presence of nature in architecture that he discerns in the stones of the Gothic cathedrals. The inherent geological record of the stones used in buildings imbues architecture with a theological capacity, a setting forth of truth and eternity. In his drawings of Mont Blanc in *Disintegration of Crystalline Rock* (1876) Viollet-le-Duc speculated on geological decay and formation. In *Style in the Technical and Tectonic Arts* (1860–1863) Semper theorized a "crystalline universe" in which the innate mathematics of geological forms gave rise to architecture. Amy Kulper argues that Semper and Viollet-le-Duc's interest is with geological processes, rather than formal attributes, for they believed in the immanence of nature's forces. They thought that the earth itself was alive and creative. Amy Catania Kulper, "Architecture's Lapidarium: On the Lives of Geological Specimens," in *Architecture in the Anthropocene: Encounters among Design, Deep Time, Science and Philosophy*, ed. Etienne Turpin (Ann Arbor, MI: Open Humanities Press, 2013), 93–95.

57. Roger Caillois, *The Mask of Medusa*, trans. G. Ordish, (New York: Clarkson N. Potter, 1964), 41. See also Caillois, *The Writing of Stones*.

58. Scholars in Critical Plant Studies have documented the neglect of vegetable life in Western thinking and established a basis for reconsidering plants. A key part of the reassessment draws on recent research that shows that plants evidence active, intentional, future-oriented movement (behavior that intrigued Darwin) and exhibit both competitive and defensive behavior. Plants may even feel pain or at least respond to danger. Plants respond to light, aromas, touch, and gravity. See Michael Marder, *Plant-Thinking: A Philosophy of Vegetal Life* (New York: Columbia University Press, 2013); Richard Doyle, *Darwin's Pharmacy: Sex, Plants, and the Noosphere* (Seattle: University of Washington Press, 2013); Matthew Hall, *Plants as Persons* (Albany: State University of New York Press, 2010); Elaine Miller, *The Vegetative Soul* (Albany: State University of New York Press, 2002); Eduardo Kohn, *How Forests Think: Toward an Anthropology beyond the Human* (Berkeley: University of California Press, 2013); Timothy Morton, *Ecology without Nature: Rethinking Environmental Aesthetics* (Cambridge, MA: Harvard University Press, 2009); Claire Colebrook,

Death of the Post-Human: Essays on Extinction, (Ann Arbor, MI: Open Humanities Press, 2013); Jeffrey Jerome Cohen, ed., *Animal, Vegetable, Mineral: Ethics and Objects* (Washington, DC: Punctum Books, 2012); Robert Mitchell, *Experimental Life: Vitalism in Romantic Science and Literature* (Baltimore, MD: Johns Hopkins University Press, 2013).

59. Daniel Chamovitz, *What a Plant Knows* (New York: Scientific American, 2013), 6, 137–138, 141.

60. Elizabeth Ellsworth and Jamie Kruse, "Introduction," in Elizabeth Ellsworth, *Making the Geologic Now: Responses to Material Conditions of Contemporary Life* (Brooklyn, NY: Punctum Books, 2013), 17.

61. To achieve LEED gold certification the Pharmaceutical Sciences Building reduced its water usage by 20 percent, used 17 percent recycled materials, provided facilities for cyclists, maximized daylight to interior spaces, and diverted 89 percent of its construction waste from landfill.

9 Floating

1. Jonathan Massey, "Ornament and Decoration," in *The Handbook of Interior Architecture and Design*, Part 3.1 *Atmospheric Conditions of the Interior*, ed. Graeme Brooker and Lois Weinthal (London: Bloomsbury, 2013), 497.

2. Gilles Deleuze, *Essays Critical and Clinical*, trans. Daniel Smith and Michael Greco (London: Verso, 1998).

3. Gilles Deleuze and Félix Guattari, *A Thousand Plateaus*, trans. Brian Massumi (Minneapolis: University of Minnesota Press, 1987), 112.

4. Correa worked with RMJM Hillier laboratory and clinical design architects and the Portuguese firm Consiste on the project. Correa was no novice when it came to the design of research facilities. He had previously designed the MIT Brain and Cognitive Sciences complex in Boston, the JN Centre for Advanced Scientific Research at Bangalore and the Inter-University Centre for Astronomy and Astrophysics at Pune. RMJM Hillier is an expert in the design of translational cancer facilities. Former projects include the Cancer Institute of New Jersey, the University Hospital Cancer Center at The University of Medicine and Dentistry of New Jersey (UMDNJ), the University of South Alabama Cancer Research Institute, the Louisiana Cancer Research Institute, and the Puerto Rico Cancer Center.

5. Deleuze and Guattari, *A Thousand Plateaus*, 112.

6. Charles Correa, "Inauguration Speech," 2010, repr. in Prasad Shetty, "Interview," *ARTIndia: The Art News Magazine of India*, accessed June 1, 2015, http://www.artindiamag.com/quarter01_01_11 /interviewPrasadShetty01_01_11.html.

7. João Silviera Botelho cited in Vernon Mays, "Champalimaud Centre for the Unknown," *Architect*, April 2011, 94.

8. The Champalimaud Foundation is a private organization established at the behest of the late Portuguese industrialist and entrepreneur, António de Sommer Champalimaud. The foundation is

composed of a board of directors, a general council, and a scientific committee. The board of directors is currently comprised of Leonor Beleza (former Portuguese minister of health and vice president of the Portuguese parliament), António Horta-Osório (chief executive [designate] of Lloyd's Banking Group), and João Botelho (former head of cabinet in two successive Portuguese governments). The general council is composed of equally eminent figures including one former president of Portugal and one of Brazil. The scientific committee is composed of internationally renowned scientists including two Nobel Prize laureates, one of whom, James Watson, chairs the committee.

9. "History," Champalimaud Foundation, http://www.fchampalimaud.org/en/the-foundation /history/.

10. João Botelho in Sankalp Meshram's short film "Into the Unknown," accessed May 18, 2015, https://www.youtube.com/watch?v=mVdcNRNPQa4.

11. Charles Correa, "Where Art and Science Meet," *ART India*, January 1, 2011, http://www .artindiamag.com/quarter01_01_11/interviewPrasadShetty01_01_11.html.

12. Champalimaud Foundation, "Public Area." accessed November 13, 2018, http://first .fchampalimaud.org/en/the-foundation/champalimaud-centre-unknown/espacos-de-fruicao -publica/.

13. Paul Goldberger, "Imitation that Doesn't Flatter," *New York Times*, April 28, 1996, http:// www.nytimes.com/1996/04/28/arts/architecture-view-imitation-that-doesn-t-flatter.html ?pagewanted=1.

14. Charles Correa in "Champalimaud Center for the Unknown," *e-architect*, March 6, 2014, http://www.e-architect.co.uk/portugal/champalimaud-foundation.

15. Correa, in Meshram, "Into the Unknown."

16. Claude Lévi-Strauss, *Introduction to the Work of Marcel Mauss*, trans. Felicity Baker (London: Routledge and Kegan Paul, 1987), 1.

17. Marcel Mauss, *A General Theory of Magic*, trans. Robert Brain (London: Routledge, 2001), 132. See also Lévi-Strauss, *Introduction to the Work of Marcel Mauss*, 34.

18. Lévi-Strauss, *Introduction to the Work of Marcel Mauss*, 63.

19. Ferdinand de Saussure, *Course in General Linguistics*, trans. Roy Harris, ed. Charles Bally and Albert Sechehaye (La Salle, IL: Open Court, 1983).

20. Jacques Lacan, *Seminar III, The Psychoses*, trans. R. Grigg (London: Routledge/Norton, 1993), 119–120.

21. Roland Barthes, "The Eiffel Tower," in *The Eiffel Tower and Other Mythologies*, trans. Richard Howard (Berkeley: University of California Press, 1997), 13.

22. Roland Barthes, "The Eiffel Tower," 4.

23. Roland Barthes, "Rhetoric of the Image," in *Image-Music-Text*, trans. Stephen Heath (London: Fontana, 1977), 39.

24. Chris L. Smith and Sandra Kaji-O'Grady, "Exaptive Translations between Biology and Architecture," *Architectural Research Quarterly (ARQ)* 8, no. 2 (2014): 155–166.

25. Barthes, *Image-Music-Text*, 31.

26. Charles Correa, "Museums: An Alternate Typology," *Daedalus* 128, no. 3 (1999): 328.

27. Correa, "Museums," 331.

28. Deleuze and Guattari, *A Thousand Plateaus*, 112.

29. Deleuze and Guattari, *A Thousand Plateaus*, 112.

30. Chris L. Smith, *Bare Architecture: A Schizoanalysis* (London: Bloomsbury, 2017), 108–110.

31. Daniel W. Smith, "Introduction: 'A Life of Pure Immanence': Deleuze's 'Critique et Clinique' Project," in Deleuze, *Essays Critical and Clinical*, trans. Daniel W. Smith and Michael A. Greco (Minneapolis: University of Minnesota Press, 1997), xvi.

32. Lévi-Strauss, *Introduction to the Work of Marcel Mauss*, 63.

33. Deleuze, *Essays Critical and Clinical*, 4.

34. Charles Correa, "The Blessings of the Sky," in *Charles Correa* (London: Thames & Hudson, 1996), 20.

35. Correa, "The Blessings of the Sky," 20.

36. Gilles Deleuze, *Cinema 1: The Movement-Image*, trans. Hugh Tomlinson and Barbara Habberjam (Minnesota, MN: Althone Press, 1986) 51.

37. Lévi-Strauss, *Introduction to the Work of Marcel Mauss*, 63.

38. Correa cited in Rick Rubens, "Champalimaud Centre," accessed June 18, 2015, http://www.rickrubens.com/Champalimaud.htm.

39. Correa cited in Rubens, "Champalimaud Centre."

40. Christian Norberg-Schulz, *Genus Loci: Toward a Phenomenology of Architecture* (New York: Rizzoli, 1991).

41. Norberg-Schulz, *Genius Loci*, 50.

42. Norberg-Schulz, *Genius Loci*, 6.

43. Correa, "The Blessings of the Sky," 28.

44. Barthes, "The Eiffel Tower."

45. Barthes, "The Eiffel Tower."

46. Charles Correa, "The Public, the Private and the Sacred," in *A Place in the Shade: The New Landscape and Other Essays* (Berlin: Hatje Cantz, 2012).

47. Correa, "The Blessings of the Sky," 27.

48. Correa, "The Blessings of the Sky," 18.

49. Correa, "The Blessings of the Sky," 27.

50. Deleuze and Félix Guattari, *A Thousand Plateaus*, 112.

51. Correa, "The Public, the Private and the Sacred."

52. Correa cited in Rubens, "Champalimaud Centre."

53. Deleuze and Guattari, *A Thousand Plateaus*, 112.

54. Lévi-Strauss, *Introduction to the Work of Marcel Mauss*, 64.

10 Symbiosis

1. The scenario is fictitious. Only one of us visited the campus on October 4 and 5, 2014, enjoying two tours—the monthly public tour led by a postdoctoral fellow and on the following day a private tour guided by Randal Jones, the long-standing campus Facilities Manager.

2. Derek H. Alderman, "Place, Naming and the Interpretation of Cultural Landscapes," in *The Ashgate Research Companion to Heritage and Identity*, ed. Brian Graham and Peter Howard (Farnham, UK: Ashgate, 2008), 199.

3. As a private, not-for-profit organization, the CSHL must comply with US laws related to tax-exemption and disclose its income, including grants, expenditure, any conflicts of interest members may have and the compensation of senior managers. Years 2011, 2012, 2013, 2014, 2015, and 2016 are publicly accessible for free, access to previous years are at a cost of USD 125. Cold Spring Harbor Laboratory, "Return of Organization Exempt from Income Tax: Schedule O," (Form 990, Department of the Treasury, Internal Revenue Service, 2012, https://www.guidestar.org/profile/11-2013303).

4. Cold Spring Harbor Laboratory, *2017 Annual Report*, https://www.cshl.edu/about-us/annual-reports/.

5. Antonio Regalado, "A Hedge-Fund Titan's Millions Stir Up Research into Autism," *Wall Street Journal*, December 15, 2005, https://www.wsj.com/articles/SB113461705596323156.

6. Jan A. Witkowski and John R. Inglis, *Davenport's Dream: 21st Century Reflections on Heredity and Eugenics* (New York: CSHL Press, 2008).

7. Mary Harriman's philanthropy saw the establishment of the Palisades Interstate Park, helped save Yosemite Valley, endowed a chair in Forestry at Yale, and supported several artists. Her support of the Eugenics Record Office is considered a blot on her reputation. Her son, William, became a New York Governor; his second wife, Marie Norton, had divorced Cornelius Vanderbilt Whitney immediately prior to marriage with William; his third wife, Pamela Churchill Harriman, had been married to the son of Winston Churchill.

8. Elizabeth L. Watson, *Houses for Science: A Pictorial History of Cold Spring Harbor Laboratory* (New York: Cold Spring Harbor Laboratory Press, 1991), 71.

9. Davenport founded the International Federation of Eugenics Organizations (IFEO) in 1925, with Eugen Fischer as chairman of the Commission on Bastardization and Miscegenation (1927). He was a leading advocate for social interventions to "improve" the American population, including sterilization of the mentally ill and policies against miscegenation and immigration. Davenport made blatantly racist claims unsupported by scientific evidence; for example, "One often sees in mulattoes an ambition and push combined with intellectual inadequacy which makes the unhappy hybrid dissatisfied with his lot and a nuisance to others" (Charles B. Davenport, "The Effects of Race Intermingling," *Proceedings of the American Philosophical Society* 56, no. 4 [1917]: 366–367).

10. Charles Darwin, *On the Origin of Species by Means of Natural Selection, or the Preservation of Favoured Races in the Struggle for Life* (1859), intro. Ernst Mayr, facs. ed. (Cambridge, MA: Harvard University Press, 1966), final paragraph. See also Stanley Hyman, *The Tangled Bank: Darwin, Marx, Frazer and Freud as Imaginative Writers* (New York: Atheneum, 1962).

11. For a detailed history of the definition of symbiosis, see chapter 1 of Angela E. Douglas, *The Symbiotic Habit* (Princeton, NJ: Princeton University Press, 2010).

12. Peter Galison, "Trading with the Enemy," in *Trading Zones and Interactional Expertise*, ed. Michael E. Gorman (Cambridge, MA: MIT Press, 2010), 32.

13. Peter Corning, *The Synergism Hypothesis: A Theory of Progressive Evolution* (New York: McGraw Hill, 1983).

14. Susan Mazur, "Lynn Margulis: Intimacy of Strangers and Natural Selection (interview)," *Scoop: Susan Mazur*, March 16, 2009, www.suzanmazur.com/?p=195.

15. Christie's, "Sale 11122, Lot 1, Dr. James D. Watson's Nobel Medal and Related Papers," December 4, 2014, New York, http://www.christies.com/lotfinder/books-manuscripts/watson -james-dewey-nobel-prize-medal-in-5857953-details.aspx.

16. Nobelprize.org, "The Nobel Prize in Physiology or Medicine 1962," *Nobel Media AB*, November 3, 2016, http://www.nobelprize.org/nobel_prizes/medicine/laureates/1962/.

17. Milmo Cahal, "Fury at DNA Pioneer's Theory: Africans Are Less Intelligent than Westerners," *Independent*, October 17, 2013.

18. Charlotte Hunt-Grubbe, "The Elementary DNA of Dear Dr. Watson," *Times Online*, October 14, 2007, http://www.thesundaytimes.co.uk/sto/culture/books/article73186.ece.

19. Keith Perry, "James Watson Selling Nobel Prize 'Because No-One Wants to Admit I Exist,'" *Telegraph*, November 28, 2014, http://www.telegraph.co.uk/news/science/11261872/James-Watson -selling-Nobel-prize-because-no-one-wants-to-admit-I-exist.html.

20. Ian Sample, "Billionaire Bought James Watson's Nobel Prize Medal in Order to Return It," *Guardian*, December 10, 2014, https://www.theguardian.com/science/2014/dec/09/russian -billionaire-usmanov-james-watson-nobel-prize-return-scientist.

21. Cold Spring Harbor Laboratory, "Return of Organization Exempt from Income Tax."

22. Sample, "Billionaire Bought James Watson's Nobel Prize Medal."

23. Michelle Speidel, "The Parasitic Host: Symbiosis Contra Neo-Darwinism," *Pli* 9 (2000): 137.

24. Slavoj Žižek, "We Don't Want the Charity of Rich Capitalists," *ABC Religion and Ethics*, August 14, 2012, http://www.abc.net.au/religion/articles/2012/08/14/3567719.htm.

25. Patricia Mooney Nickel, "Haute Philanthropy: Luxury, Benevolence and Value," *Luxury* 2, no. 2 (2016): 13.

26. Nickel, "Haute Philanthropy," 26.

27. Carol Strickland, "Watson Relinquishes Major Role at Lab," *New York Times*, March 21, 1993, http://www.nytimes.com/1993/03/21/nyregion/watson-relinquishes-major-role-at-lab.html?pagewanted=all.

28. Robert C. Hughes, *Cold Spring Harbor* (Charleston, SC: Arcadia Publishing, 2014), 7.

29. Watson, *Houses for Science*, 38.

30. Watson, *Houses for Science*, 42.

31. Watson, *Houses for Science*, 169.

32. Victor K. McElheny, *Watson and DNA: Making a Scientific Revolution* (New York: Perseus/Wiley, 2003), 164.

33. James Watson, *Avoid Boring People: Lessons from a Life in Science* (Oxford: Oxford University Press, 2007), 295–296.

34. Watson, *Avoid Boring People,* 312–313.

35. Watson, *Avoid Boring People,* 313.

36. McElheny, *Watson and DNA,* 169.

37. James Watson quoted in Enid Nemy, "Lita Hazen, Patron of Sciences, Dies at 85," *New York Times*, October 3, 1995. http://www.nytimes.com/1995/10/03/obituaries/lita-hazen-patron-of-sciences-dies-at-85.html?pagewanted=all.

38. Wikipedia, "Cold Spring Harbor, New York," accessed February 15, 2015, http://en.wikipedia.org/wiki/Cold_Spring_Harbor,_New_York.

39. Laurel Hollow was ranked eleventh by *Worth* magazine in 1996 using 1990 census figures on household incomes and housing values. Businessweek.com ranked Laurel Hollow the twenty-fourth wealthiest small town in America in 2011 (defined as less than 10,000 people) using only median home value data, while 24/7 Wall Street ranked it 9th in 2017 based on household income and home value. Census bureau figures for Laurel Hollow show an average household income of USD 293,345 for 2006. John Rather, "'Wealthiest? Who, Us? Long Island Villagers Exclaim," *New York Times*, August 18, 1996, https://www.nytimes.com/1996/08/18/ … /wealthiest-who-us-li-villagers-exclaim.html; "America's 50 Most Expensive Small Towns

2011," *Bloomberg*, January 22, 2011, https://www.bloomberg.com/news/photo-essays/2011-01-21/americas-50-most-expensive-small-towns-2011; Evan Comen, "America's Richest (and Poorest) Towns, *24/7 Wall Street*, May 8, 2017, https://247wallst.com/special-report/2017/05/08/americas-richest-and-poorest-towns-2/5/.

40. The nine towns are Brookville, Lloyd Harbor, Munsey Park (Manhasset), Muttontown, North Hills, Old Westbury, Oyster Bay Cove, Roslyn Estates, and Sands Point. See also Prashant Gopal, "Where the Rich Still Live," *Bloomberg*, March 17, 2009, http://www.bloomberg.com/bw/lifestyle/content/mar2009/bw20090317_499218.htm.

41. Cold Spring Harbor Laboratory, "Honor Roll of Donors," report (New York, 2013), http://www.cshl.edu/images/stories/about_us/development/2013_Donor-HonorRoll.pdf.

42. Edwin Durgy, "Billionaires for the Cure: James Simon, Autism," *Forbes*, March 14, 2012, accessed February15, 2015, http://www.forbes.com/pictures/eilm45mll/james-simons-autism/.

43. Biondi studied at Yale, Nicholls at Amherst College. They met while graduate students at Harvard Business School, married in 1997, and have three children, Serena, Carter and Joy.

44. "Weddings; Jamie Nicholls, Ottavio Biondi Jr.," *New York Times*, February 2, 1997, http://www.nytimes.com/1997/02/02/style/jamie-nicholls-ottavio-biondi-jr.html.

45. "Deaths: Nicholls, Richard Hall," *New York Times*, March 22, 2009, http://query.nytimes.com/gst/fullpage.html?res=9900EEDD133AF931A15750C0A96F9C8B63.

46. "Deaths: Nicholls, Richard Hall."

47. "Deaths: Nicholls, Richard Hall."

48. For example, the Weill Cornell's Belfer Building (2014) received USD 100 million from Bob and Renée Belfer, along with 154 gifts of USD 1 million or more, bringing the total number of donations to USD 400 million. "Hope for All," *New York Social Diary*, November 17, 2011, http://www.newyorksocialdiary.com/party-pictures/2011/hope-for-all.

49. Weill Cornell Newsroom, "Weill Cornell Opens Its Transformative Belfer Research Building, Empowering Scientists to Speed Discoveries to Patients," press release, January 31, 2014, http://weill.cornell.edu/news/pr/2014/01/weill-cornell-opens-its-transformative-belfer-research-building-empowering-scientists-to-speed-disco.html.

50. James Watson quoted in McElheny, *Watson and DNA*, 8.

51. Watson, *Avoid Boring People*, 286.

52. Cold Spring Harbor Laboratory, "Double Helix Medals Dinner," accessed February 12, 2015, http://www.cshl.edu/DHMD.

53. "Cold Spring Harbor Laboratory's Double Helix Medals Dinner," accessed February 12, 2015, http://www.patrickmcmullan.com/site/event_detail.aspx?eid=27932.

54. This point is nowhere better made than in fiction, where it is free of any self-serving institutional agenda. Jeffrey Eugenides's novel *The Marriage Plot* is partly set in the Cold Spring Harbor

Laboratory in 1982, where it appears under the guise of Pilgrim Lake. The margins of the book's plot are populated by real figures; James Watson appears as the figure of Dr. Malkiel and Barbara McClintock as Diane MacGregor.

55. Centerbrook Architects and Planners, "Computational Neuroscience Laboratory," accessed February 12, 2015, http://www.centerbrook.com/project/cold_spring_harbor_laboratory_computational_neuroscience_laboratory.

56. Watson, *Houses for Science*, 315.

57. Watson, *Houses for Science*, 315.

58. Elizabeth Watson, *Grounds for Knowledge: A Guide to Cold Spring Harbor Laboratory's Landscapes and Buildings* (Cold Spring Harbor: Cold Spring Harbor Laboratory Press, 2008), 127.

59. Cold Spring Harbor Laboratory, "New Research Buildings Open at Cold Spring Harbor Laboratory," in "News and Features," accessed February 12, 2015, http://www.cshl.edu/news-a-features/new-research-buildings-open-at-cold-spring-harbor-laboratory.html.

60. Centerbrook Architects and Planners, "Hillside Research Campus: Cold Spring Harbor Laboratory," accessed February 12, 2015, https://centerbrook.com/project/cold_spring_harbor_laboratory_hillside_research_campus.

61. Centerbrook Architects and Planners, "Marks Laboratory for Neuron Imaging: Cold Spring Harbor Laboratory," accessed February 12, 2015, http://www.centerbrook.com/project/cold_spring_harbor_laboratory_marks_laboratory_for_neuron_imaging.

62. Randal Jones, email correspondence with Sandra Kaji-O'Grady, April 28, 2015.

63. Jim Childress, "Watson and Grover Pack the House," *The Millrace*, December 28, 2010, http://centerbrook.com/blog/2010/12/watson-and-grover-pack-the-house/.

64. J. Alex Tarquinio, "Long Island Laboratory's Expansion Hides in (and under) Six Buildings," *New York Times*, June 23, 2009, http://www.nytimes.com/2009/06/24/realestate/commercial/24lab.html.

65. James Childress, "Cold Spring Harbor Upgrade Creates Research Village," *Laboratory Design*, August 16, 2010, http://www.labdesignnews.com/articles/2010/08/cold-spring-harbor-upgrade-creates-research-village.

66. Childress, "Cold Spring Harbor Upgrade Creates Research Village."

67. Tarquinio, "Long Island Laboratory's Expansion Hides in (and under) Six Buildings."

68. Centerbrook Architects and Planners, "Firm Profile," accessed February 9, 2015, http://www.centerbrook.com/about/firm.

69. Jorge Otero-Pailos, *Architecture's Historical Turn: Phenomenology and the Rise of the Postmodern* (Minneapolis: University of Minnesota Press, 2010).

70. Anthony Vidler, *The Architectural Uncanny: Essays in the Modern Unhomely* (Cambridge, MA: MIT Press, 1992), 219.

71. Jamie Nicholls and her husband, Francis Biondi, bought their home, originally built in 1936, on 5.2 acres for USD 8 million in 2005. Long Island Profiles: Ten Year Real Estate Sales History, accessed February 11, 2015, http://listings.findthehome.com/l/72345590/208-Cleft-Rd -Mill-Neck-NY-11765. Francis Biondi was ranked 387th in the United States on Forbes's 2016 list of "The World's Billionaires," with a net worth of USD 1.55 billion. See Real Time Net Worth, "The World's Billionaires: The Richest People on the Planet 2016–1121 O. Francis Biondi," *Forbes*, December 2, 2016, http://www.forbes.com/profile/o-francis-biondi.

72. Watson, *Houses for Science*, x.

73. Watson, *Houses for Science*, x.

74. Svetlana Boym, *The Future of Nostalgia*, (New York: Basic Books, 2001), 350–352.

11 Aggrandizement

1. Bradley J. Fikes, "Biotech Cluster Gets Serious Star Power," *San Diego Union-Tribune*, October 5, 2013, http://www.sandiegouniontribune.com/news/2013/oct/05/craig-venter-biotech-science -institute-jolla/.

2. Sam Keane, *The Violinist's Thumb: And Other Lost Tales of Love, War, and Genius, as Written by Our Genetic Code* (New York: Little, Brown and Company, 2012).

3. John Sulston and Georgina Ferry, *The Common Thread: A Story of Science, Politics, Ethics and the Human Genome* (London: Corgi, 2003), 176.

4. Craig Venter, "The Relationship with James Watson," Cold Spring Harbor Laboratory: DNA Learning Centre, *DNAi* video, 0:53, accessed 19 March 19, 2019. http://www.dnalc.org/view /15357-The-relationship-with-James-Watson-Craig-Venter.html.

5. Jeremy Rifkin quoted in Meredith Wadman, "Biology's Bad Boy Is Back; Craig Venter Brought Us the Human Genome. Now He Aims to Build a Life Form That Will Change the World," *Fortune Magazine*, March 8, 2004, http://archive.fortune.com/magazines/fortune/fortune_archive/2004 /03/08/363705/index.htm.

6. Craig Venter drew a salary of USD 708,462 from the JCVI in 2013. Venter's wife, Heather Kowalski, was paid USD 120,000 for "public relations services."

7. It is somewhat bizarre that Venter's subsidiary, Genovia, bears the same name as the fictional European principality of the teenage novel series *The Princess Diaries* (2000).

8. The description of Genovia Bio's mission was originally taken from the Synthetic Genomics website: accessed 4 August 4, 2015, http://web.archive.org/web/20050924142325 /http://syntheticgenomics.com/about.htm. The Synthetic Genomics site was dramatically rede-signed in 2016 and now details projects and ambitions, but no longer links these to the various subsidiaries. A longer description can be found on Wikipedia, accessed March 19, 2019, https://en .wikipedia.org/wiki/Synthetic_Genomicshttps://en.wikipedia.org/wiki/Synthetic_Genomics.

9. Heather Kowalski, "Synthetic Genomics Inc. and Plenus, S.A de C.V Form New Sustainable Agriculture Company, Agradis, Inc. to Develop and Commercialise Products Using Genomic Technologies," press release, October 24, 2011, http://www.prnewswire.com/news-releases /synthetic-genomics-inc-and-plenus-sa-de-cv-form-new-sustainable-agriculture-company-agradis -inc-to-develop-and-commercialize-products-using-genomic-technologies-132459158.html.

10. The University of San Diego has a collaborative research agreement with Human Longevity that allows the private company to access Moores Cancer Center patient data.

11. Human Longevity, Inc., accessed August 21, 2015, http://www.humanlongevity.com/.

12. *The San Diego Union-Tribune*, reporting the construction of the JCVI, stated that "local science leaders anticipate that, as with polio vaccine pioneer Salk, the intellectual heft of Venter and his researchers will elevate San Diego even more as a world center of science." Fikes, "Biotech Cluster Gets Serious Star Power."

13. Richard Ebright, of the Howard Hughes Medical Institute, Rutgers University, quoted in Wadman, "Biology's Bad Boy Is Back."

14. "James Watson: Craig Venter Is a Great Marketer," *YouTube* video, 3:41, posted by "Big Think," June 27, 2011, https://www.youtube.com/watch?v=sxTrdLtAIY8.

15. Georgina Ferry, "Learning the Lessons of Life," *Guardian*, October 27, 2007, http://www .theguardian.com/books/2007/oct/27/featuresreviews.guardianreview6.

16. Steven Shapin, "I'm a Surfer," *London Review of Books* 30, no. 6 (2008): 5–8, http://www.lrb.co .uk/v30/n06/steven-shapin/im-a-surfer.

17. Wadman, "Biology's Bad Boy Is Back."

18. Sara Lin, "Craig Venter's Hangout," *Wall Street Journal*, March 12, 2010, http://www.wsj.com /articles/SB10001424052748704548604575098131101212908.

19. Water samples across a several-thousand kilometer transect from the North Atlantic through the Panama Canal and ending in the South Pacific enabled a metagenomic study of the marine planktonic microbiota, yielding an extensive dataset consisting of 7.7 million sequencing reads (6.3 billion bp). D. B. Rusch, A. L. Halpern, G. Sutton, et al., "The Sorcerer II Global Ocean Sampling Expedition: Northwest Atlantic through Eastern Tropical Pacific," *PLoS Biology* 5, no. 3 (2007): e77, http://journals.plos.org/plosbiology/article?id=10.1371/journal.pbio.0050077.

20. Other figures in the Jaeger LeCoultre campaign are Argentine polo player Eduardo Novillo Astrada and actress, screenwriter, and producer Carmen Chaplin.

21. The apparent intention is to portray the three individuals in the campaign in the context in which they made their names and fortunes, or, as claimed by *The Jewelery Editor*, an online luxury magazine with an alleged monthly global reach of 2.8 million, we are invited to "step into the worlds of extraordinary people." The forty-three-year old Novillo Astrada—a long standing "ambassador" for the company—is pictured in riding gear leading a horse into a cool, dark stable,

the sun behind him hot and bright. Chaplin, improbably gamine at age thirty-eight, is shown captivated by a silent movie, possibly starring her grandfather, Charlie Chaplin. Neither Chaplin nor Novilla Astrada has changed the world through their work as Venter has, but their dark good looks, relative youth, and privileged family backgrounds—Novilla Astrada's father was also a world renowned polo player and his wife, Astrid Munoz, is a former model and photographer for Jaeger Le Coultre—lends each an ease and grace before the camera that eludes Venter. "Jaeger-LeCoultre Broadens its Horizons with a Sophisticated New Advertising Campaign," *The Jewelery Editor*, May 10, 2015, http://www.thejewelleryeditor.com/watches/jaeger-le-coultre-watches-new -advertising-campaign/.

22. The Jaeger-LeCoultre campaign was conceived by advertising agency, Agency DDB, with creative direction from Christian Vince. Venter's casting was managed by TMA Lux, the boutique division of The Marketing Arm, designed to serve luxury brands. The films were produced by Julien Pasquier from the French company Standard Films, with Matias Boucard as director of photography. Post production they were edited by Mark Maborough, colored by Mathieu Caplanne, and with sound design by THE.

23. "Jaeger-LeCoultre Broadens Its Horizons."

24. "Our Makers: Interview Craig Venter – The Greatest Moments of Our Time by Jaeger-LeCoultre," accessed March 18, 2019, https://www.youtube.com/watch?v=n_ti6ThwVxo.

25. Venter has long kept poodles. A previous pet of Venter and his then wife, geneticist Claire Fraser, was a standard poodle named Shadow. Shadow was the first dog to have a draft sequence of its genome mapped. Venter got Darwin when the dog was three months old, and "we bonded immediately." He explains, "The intellectual capacity and observational learning that Darwin has and other poodles have is truly stunning. He's a great companion for both of us. He's traveled around the world with us. We sailed across Italy and Greece and up into Turkey. He's seen an awful lot in his six years" (quoted in Bradley J. Fikes, "Groundbreaking Quest to Improve Lives," *San Diego Tribune*, July 1, 2016, http://www.sandiegouniontribune.com/news/sdut-craig-venter -genomics-human-genome-synthetic-2016jul01-story.html).

26. In an interview in 2013, Venter says, "I think the work that we have done with the first genome in history, the human genome and with the first synthetic cell is certainly of the world calibre that obviously earns big prizes. Nobel prizes are very special prizes and it would be great to get one" (quoted in Zoë Corbyn, "Craig Venter: 'This Isn't a Fantasy Look at the Future. We Are Doing the Future,'" *Guardian*, October 13, 2013, https://www.theguardian.com/science/2013/oct /13/craig-ventner-mars).

27. BD+C Staff, "Ranked: Top Science and Technology Sector AEC Firms [2014 Giants 300 Report]," August 21, 2014, http://www.bdcnetwork.com/ranked-top-science-and-technology -sector-aec-firms-2014-giants-300-report.

28. J Craig Venter Institute, "Return of Organization Exempt from Income Tax" (Form 990, Department of the Treasury, Internal Revenue Service, 2013), 8, accessed August 21, 2015, http:// www.guidestar.org/FinDocuments/2013/521/842/2013-521842938-0aefb777–9.pdf.

29. BD+C Staff, "Ranked: Top Science and Technology Sector AEC Firms."

30. Marsha King, "In This Space at This Time—ZGF's Organic Style Gives Birth to Buildings That Fit," *Seattle Times*, July 28, 1991, http://community.seattletimes.nwsource.com/archive/?date=19910728&slug=1296834.

31. Monographs such as Mildred Schmertz and Deborah Dietsch, *Zimmer Gunsul Frasca: Building Community* (Rockport, Mass: Rockport Publishers, 1995) and Robert Frasca and Joseph Giovannini, *Future Tense: Zimmer Gunsul Frasca* (South Pasadena, CA: Balcony Press, 2008) are essentially self-authored, coffee-table fare paid for by the company. Balcony Press creates custom books to "showcase exceptional design work, enhance brand image and commemorate occasions". Balcony Press homepage, accessed March 19, 2019, http://www.balconypress.com/html/request.html.

32. Ted Hyman, quoted in ZGF Architects LLP, "J. Craig Venter Institute" (online portfolio). p. 9, accessed August 11, 2015, http://issuu.com/zgfarchitectsllp/docs/j._craig_venter_institute?e=5145747/7966125.

33. Craig Venter, quoted in Fikes, "Groundbreaking Quest to Improve Lives."

34. McCarthy Building Companies, "J. Craig Venter Institute on USCD Campus Expected to Be Only Net Zero Energy Biological Laboratory in the World," press release, March 17, 2013, http://www.mccarthy.com/insights/j-craig-venter-institute-ucsd-campus-expected-be-only-net-zero-energy-biological-laboratory.

35. Hyman, quoted in ZGF Architects LLP, "J. Craig Venter Institute," 9.

36. ZGF Architects LLP, "J. Craig Venter Institute," 9.

37. Craig Venter, quoted in ZGF Architects LLP, "J. Craig Venter Institute," 7

38. Craig Venter, *A Life Decoded: My Genome, My Life* (New York: Penguin, 2007), 334.

39. Bruno Latour, "Give Me a Laboratory and I Will Raise the World," in *Science Observed: Perspectives on the Social Study of Science*, ed. Karin D. Knorr-Cetina and Michael Mulkay (London: Sage, 1983), 152.

40. Venter, *A Life Decoded*, 334.

41. Venter, *A Life Decoded*, 348.

42. Janet Howard, "Maverick of Science Finds His Match," UC San Diego News Center, February 13, 2013, http://ucsdnews.ucsd.edu/feature/maverick_of_science_finds_his_match.

43. The examples here are from Paul Hawken, ed., *Drawdown: The Most Comprehensive Plan Ever Proposed to Reverse Global Warming* (New York: Penguin, 2017).

44. Lin, "Craig Venter's Hangout."

45. Lin, "Craig Venter's Hangout."

46. Lin, "Craig Venter's Hangout."

47. Founded in 1979, Nexus Properties, Inc., is a developer of corporate facilities, biotech laboratories, and flexible research and development properties throughout California, Washington, and North Carolina. Nexus has constructed numerous multi-tenanted and build-to-suit laboratory facilities for life science companies such as Ligand Pharmaceuticals, Johnson & Johnson, Amgen, and Neurocrine Biosciences, as well as incubator facilities for small young biotech companies.

48. Wil S. Hylton, "Craig Venter's Bugs Might Save the World," *New York Times Magazine*, May 30, 2012, http://www.nytimes.com/2012/06/03/magazine/craig-venters-bugs-might-save-the-world .html?_r=0.

49. Lin, "Craig Venter's Hangout."

50. Employees offer mixed reviews of the JCVI on the Glassdoor website (a site for employees to anonymously review their past or present places of work). Laboratory facilities, the new building, and the opportunity to work with bright colleagues are highly valued, but the complaints about management style across forty-eight reviews are consistent and damning of Venter's preference to install loyal and long-serving scientist colleagues into management positions. Anonymous, "J Craig Venter Institute Reviews," Glassdoor, accessed August 15, 2015, https://www.glassdoor.com .au/Reviews/J-Craig-Venter-Institute-Reviews-E111330_P2.htm.

51. "Human Longevity Acquires Stem Cell Banking Business Unit from Cologne," GenomeWeb, January 28, 2016, https://www.genomeweb.com/business-news/human-longevity-acquires-stem -cell-banking-business-unit-celgene.

52. Juan Enriquez's books include: *As the Future Catches You: How Genomics & Other Forces Are Changing Your Life, Work, Health & Wealth* (2001), *Homo Evolutis* (2010), and *Evolving Ourselves: How Unnatural Selection and Nonrandom Mutation Are Changing Life on Earth* (2015).

53. Andrew Pollack, "His Corporate Strategy: The Scientific Method," *New York Times*, September 4, 2010, http://www.nytimes.com/2010/09/05/business/05venter.html?_r=0.

54. Anthropologist Paul Rabinow's book, co-authored with Talia Dan-Cohen, *A Machine to Make a Future: Biotech Chronicles* (Princeton, NJ: Princeton University Press, 2005), is based on Celera.

55. Lewis D. Solomon, *Synthetic Biology: Science, Business, and Policy* (Piscataway, NJ: Transaction Publishers, 2011), 104.

56. Craig Venter, "The Richard Dimbleby Lecture: Dr J Craig Venter—A DNA-Driven World," lecture transcript, Richard Dimbleby Lecture as broadcast on *BBC One*, December 4, 2007, http:// www.bbc.co.uk/pressoffice/pressreleases/stories/2007/12_december/05/dimbleby.shtml.

57. Craig Venter quoted in Melinda Wenner, "The Next Generation of Biofuels", *Scientific American*, March 1, 2009, http://www.scientificamerican.com/article/the-next-generation-of-biofuels/.

58. Pollack, "His Corporate Strategy."

59. Archer Daniels Midland, for example, has been the subject of several major federal lawsuits in the United States related to air pollution. The firm was also investigated with regard to lysine

price-fixing by the US Justice Department in the mid-1990s, a case that became the basis for the thriller film *The Informant* (2009).

12   Investments

1. Nigel Thrift, "Re-Inventing Invention: New Tendencies in Capitalist Commodification," *Economy and Society* 35, no. 2 (2006): 293.

2. Francis Duffy and Jack Tanis, "A Vision of the New Workplace," *Site Selection and Industrial Development* 162, no. 2 (1993): 428.

3. Maurizio Lazzarato, "Immaterial Labor," in *Radical Thought in Italy: A Potential Politics*, ed. Paolo Virno and Michael Hardt, trans. Paul Colilli and Ed Emery (Minneapolis: University of Minnesota Press, 1996), 133–147.

4. Alan McKinlay and Philip Taylor, "Power, Surveillance and Resistance: Inside the 'Factory of the Future,'" in *The New Workplace and Trade Unionism*, ed. P. Ackers, C. Smith and P. Smith (London: Routledge, 1996), 285; emphasis in original.

5. Vishal Bupta, Shailendra Singh, and Naresh Khatri, "Creativity in Research and Development Laboratories: A New Scale for Leader Behaviours," *IIMB Management Review* 25, no. 2 (2013): 66.

6. Lisa Adkins and Celia Lury, "The Labor of Identity: Performing Identities, Performing Economies," *Economy and Society* 28, no. 4 (1999): 601.

7. Adkins and Lury, "The Labor of Identity," 601.

8. Maurizio Lazzarato, *Signs and Machines: Capitalism and the Production of Subjectivity*, trans. Joshua Jordan (Los Angeles: Semiotext(e), 2014), 44.

9. McKenzie Wark, "Lazzarato and Pasolini," *Public Seminar*, June 15, 2015, http://www .publicseminar.org/2015/06/lazzarato-and-pasolini/.

10. Maurizio Lazzarato, "From Capital-Labour to Capital-Life," *Ephemera: Theory and Politics in Organization* 4, no. 3 (2004): 194.

11. Felix Guattari, "Les annees d'hiver" (2009), quoted in Maurizio Lazzarato, "'Exiting Language': Semiotic Systems and the Production of Subjectivity in Félix Guattari," in *Cognitive Architecture: From Bio-politics to Noo-politics*, trans. Eric Anglés, ed. Deborah Hauptmann and Warren Neidich (Rotterdam: 010 Publishers, 2010), 503.

12. Lazzarato, "'Exiting Language,'" 503.

13. New York City Economic Development Corporation, "Mayor Bloomberg, Speaker Silver and Alexandria Real Estate Equities Open New State-of-the-Art Science Park in Manhattan," press release, December 2, 2010, http://www.nycedc.com/press-release/mayor-bloomberg-speaker-silver -and-alexandria-real-estate-equities-open-new-state-art.

14. This may be because former directors of the US architecture firm Hillier sued Scotland-based RMJM (Robert Matthew Johnson Marshall) in 2011 after RMJM failed to pay employees an

amount agreed when the two firms merged in 2007. RMJM was "bailed out" GBP 8 million that year by the Morrison family, which includes the RMJM chief executive, Peter Morrison and his father, Chairman Sir Fraser Morrison. In 2012, three of RMJM's British subsidiaries went into receivership and were sold to the newly formed RMJM Architecture. Peter Morrison commended his team across the world for their "resilience and loyalty," but several of the leaders of the firm's US practice resigned during this period, citing their disagreement with business decisions made by the firm's Scottish directors. Grieg Cameron, "RMJM's Subsidiaries Put into Receivership," *Herald Scotland*, October 27, 2012. http://www.heraldscotland.com/business/13078551.RMJM_s _subsidiaries_put_into_receivership/.

15. Alexandria, "About Alexandria," accessed September 29, 2018, http://www.alexandrianyc .com/ersp.html.

16. Russell Hughes details this story in "The Internet of Politicised 'Things': Urbanisation, Citizenship, and the Hacking of New York 'Innovation' City," *Interstices* 16 (2016): 24–30.

17. New York City Economic Development Corporation, "Mayor Bloomberg Announces Developer to Build Largest Biotech Campus in New York City," press release, August 10, 2005, http:// www.nycedc.com/press-release/mayor-bloomberg-announces-developer-build-largest-biotech -campus-new-york-city.

18. New York City Economic Development Corporation, "Mayor Bloomberg, Governor Paterson, Speaker Silver, Eli Lilly and Company and Alexandria Real Estate Equities Announce Imclone Will Locate Its Research Headquarters at East River Science Park," press release, July 22, 2009. http://www.nycedc.com/press-release/mayor-bloomberg-governor-paterson-speaker-silver-eli -lilly-and-company-and-alexandria.

19. Steven M. Paul, Daniel Mytelka, Christopher Dunwiddle, Charles Persinger, Bernard Munos, Stacy Lindborg, and Aaron Schacht, "How to Improve R&D Productivity: The Pharmaceutical Industry's Grand Challenge," *Nature Reviews: Drug Discovery* 9 (March 2010): 208.

20. Matthew Owens, Harlem Biospace executive director, quoted in Ainsley O'Connell, "The Big Apple's Biotech Dreams Are Stuck in the Petri Dish," *Fast Company*, February 24, 2015, http:// www.fastcompany.com/3034774/new-york-biotech-startup-dreams.

21. Harlem Biospace, "Who We Are," accessed September 9, 2016, http://harlembiospace.com/.

22. Matthew Owens, quoted in O'Connell, "The Big Apple's Biotech Dreams."

23. Harlem Biospace, "Who We Are."

24. Tandon has a bachelor's degree in Electrical Engineering from Cooper Union, a master's degree in Bioelectrical Engineering from MIT, a PhD in Biomedical Engineering, and an MBA from Columbia University. Her PhD research studied electrical signaling in cardiac, skin, bone, and neural tissue.

25. Liz Welch, "How a Bone-Growing Startup Lured 66 Investors, Including Peter Thiel," *Inc. Magazine*, October 2015. http://www.inc.com/magazine/201510/liz-welch/blooming-bones.html ?cid=sf01001.

26. Zaina Awad, "Making a Living with Biology—Q&A with Nina Tandon," TEDMED, May 7, 2015. http://blog.tedmed.com/field-living-possibilities-qa-nina-tandon/.

27. Awad, "Making a Living with Biology."

28. "How to Regrow Your Own Bones," *Scientific American*, June 27, 2016, http://www.scientificamerican.com/article/how-to-regrow-your-own-bones/?WT.mc_id=SA_TW_TECH_FEAT.

29. Welch, "How a Bone-Growing Startup Lured 66 Investors."

30. EpiBone, "Our Team," accessed September 2, 2016, http://EpiBone.com/team.

31. "Nina Tandon, PhD '09 BME," Graduate Student Affairs, Columbia Engineering, accessed September 3, 2016, http://gradengineering.columbia.edu/nina-tandon-phd-09-bme.

32. "Nina and Noah," Zola Wedding Registry, August 27, 2016, https://www.zola.com/registry/ninaandnoah?pkey=affiliate&utm_medium=affiliate&utm_source=TnL5HPStwNw&utm_campaign=1.

33. Vivian Yee, "Salvaging a Long-Lasting Wood, and New York City's Past," *New York Times*, July 21, 2015, http://www.nytimes.com/2015/07/22/nyregion/salvaging-a-long-lasting-wood-and-new-york-citys-past.htm.

34. Nina Tandon, Pinterest, accessed September 3, 2016, https://au.pinterest.com/ioanaru/tandon/.

35. Keiko Morris, "Wanted: Biotech Startups in New York City: The Alexandria Center for Life Science Looks to Expand," *Wall Street Journal*, July 28, 2014, http://www.wsj.com/articles/wanted-biotech-startups-in-new-york-city-1406603189.

36. Cornell Tech Homepage, "Runway Startup Postdocs," accessed 19 March 19, 2019, https://tech.cornell.edu/programs/startup-postdocs.

37. The Eames Executive Work Chair was designed for the Time-Life Building in 1960 and today is manufactured and sold through Herman Miller; the chairs in Alexandria could be these or replicas.

38. Alexandria LaunchLabs, accessed September 5, 2016, http://www.launchlabsnyc.com/.

39. "Alexandria LaunchLabs, the Premier Life Science Startup Platform, to Open in Fall 2018 at the Alexandria Center at One Kendall Square in the Heart of East Cambridge," *PR Newswire*, April 12, 2018, https://www.prnewswire.com/news-releases/alexandria-launchlabs-the-premier-life-science-startup-platform-strategically-located-at-the-alexandria-center-for-life-science—nyc-celebrates-first-anniversary-and-announces-first-investments-through-the-alexandria-seed-capi-300665937.html.

40. Gilles Deleuze and Felix Guattari, *A Thousand Plateaus: Capitalism and Schizophrenia* [1980], trans. and foreword, Brian Massumi (London: Continuum, 2004), 237.

41. O'Connell, "The Big Apple's Biotech Dreams."

42. Jenna Foger, quoted in Christian Bautista, "Alexandria Real Estate Equities Launches Co-Working for Science Guys," *Real Estate Weekly*, June 15, 2016, http://rew-online.com/2016/06/15/alexandria-real-estate-equities-launches-co-working-for-science-guys/.

43. USD 51 million is claimed to have been raised in partnership with Accelerator Corp, a biotech investment and management firm, to attract and support biotech start-ups to launch themselves here. Investors include the Partnership Fund for New York City, Harris & Harris Group, ARCH Venture Partners, WRF Capital, and the major pharmaceutical companies already housed on the campus. Morris, "Wanted: Biotech Startups in New York City."

44. "Venture Capital Investing Exceeds $17 Billion for the First Time According to the Moneytree Report," PriceWaterhouseCoopers, July 17, 2015, http://www.pwc.com/us/en/press-releases/2015/venture-capital-investing-exceeds.html.

45. Meg Tirrell, "Biotech's Real Estate Boom," *CNBC News*, April 30, 2015, http://www.cnbc.com/2015/04/30/biotechs-real-estate-boom.html.

46. Tirrell, "Biotech's Real Estate Boom."

47. Joe Gose, "Why Rental Rates at Biotech Labs are Skyrocketing," *Investor's Business Daily*, October 15, 2015, http://www.investors.com/news/real-estate/biotech-lab-rental-rates-rising/.

48. Karen Weintraub, "Biotech Players Lead a Boom in Cambridge," *New York Times*, January 1, 2013, http://www.nytimes.com/2013/01/02/realestate/commercial/biotech-players-lead-a-boom-in-cambridge.html.

49. Roger Showley, "BioMed Ups UTC Footprint with New $110M Purchase," *San Diego Union-Tribune*, April 11, 2016, http://www.sandiegouniontribune.com/news/2016/apr/11/biomed-illumina-utc/.

50. Lazzarato, "From Capital-Labour to Capital-Life," 195.

51. Angela McRobbie, *Be Creative: Making a Living in the New Culture Industries* (London: Polity, 2016), 37.

52. McRobbie, *Be Creative*, 153.

53. Harlem BioSpace, "Applying to Harlem BioSpace: FAQ," accessed September 9, 2016. http://harlembiospace.com/apply-for-space/.

54. Tammy J. Erickson, "Straight from Hollywood: The Project-Based Workforce," *Harvard Business Review*, January 29, 2008, https://hbr.org/2008/01/straight-from-hollywood-the-pr.

55. Daniel Lee Kleinman and Stephen P. Vallas, "Science, Capitalism, and the Rise of the 'Knowledge Worker': The Changing Structure of Knowledge Production in the United States," *Theory and Society* 30, no. 4 (2001): 464.

56. Michel Foucault, Preface to Gilles Deleuze and Félix Guattari, *Anti-Oedipus*, trans. Robert Hurley, Mark Seem, and Helen R. Lane (Minneapolis: University of Minnesota Press, 1983), xiii.

57. Alan Liu, *The Laws of Cool: Knowledge Work and the Culture of Information* (Chicago: University of Chicago Press, 2004), 236.

13 Conclusion

1. Jeremy Coote, "Origins of the Afro Comb: 6,000 Years of Culture, Politics, and Identity (review)," *African Arts* 50, no. 4 (2017): 81–82.

2. Dan Howarth, "Renzo Piano Completes 'Palace of Light' for Columbia University Medical Researchers," *Dezeen*, October 25, 2016, https://www.dezeen.com/2016/10/25/renzo-piano-jerome -l-greene-science-center-columbia-university/.

3. Georges Canguilhem, *Ideology and Rationality in the History of the Life Sciences* (1977), trans. Arthur Goldhammer (Cambridge, MA: MIT Press, 1988), 13. See also Michel Foucault, *Archaeology of Knowledge*, trans. Alan Mark Sheridan-Smith (New York: Pantheon Books, 1972), 224.

4. Niles Eldredge and Stephen Jay Gould, "Punctuated Equilibria: An Alternative to Phyletic Gradualism," in *Models in Paleobiology*, ed. T. J. M. Schopf, (San Francisco: Freeman Cooper, 1972), 82–115.

5. Niles Eldredge has spent his working life explaining evolution and expanding its arguments. As curator and research paleontologist in the Department of Invertebrate Paleontology at the American Museum of Natural History since 1969, he organized the museum's acclaimed exhibit celebrating the bicentennial of Darwin's birth and the sesquicentennial of the publication of *On the Origin of Species* (1859). His research specialty is the evolution of mid-Paleozoic Phacopida trilobites, a group of extinct arthropods that lived between 543 and 245 million years ago. See Gabrielle Walker, "The Collector," *New Scientist* 179, no. 2405 (July 26, 2003): 38–41.

6. Belinda Barnet and Niles Eldredge, "Material Cultural Evolution: An Interview with Niles Eldredge," *Fibreculture Journal* 3 (2004), accessed April 20, 2015, http://three.fibreculturejournal .org/fcj-017-material-cultural-evolution-an-interview-with-niles-eldredge/.

7. "Skolkovo Institute of Science and Technology/Herzog and de Meuron," *Archdaily*, accessed November 18, 2018, https://www.archdaily.com/905951/skolkovo-institute-of-science-and -technology-herzog-and-de-meuron.

8. "Skolkovo Institute of Science and Technology/Herzog and de Meuron."

9. Wellcome Genome Campus, "Future Plans," accessed November 16, 2018, https://www .wellcomegenomecampus.org/aboutus/futureplans/.

Selected Bibliography

The selected bibliography consists of scholarly texts that shaped the arguments made in *LabOratory*. Other sources that are not specifically referenced in the text but helped to frame our thinking are also included. As discussed in chapter 2, we drew on a much broader set of sources to tell the story of the contemporary laboratory, such as the news releases generated by architectural offices that are disseminated through online channels such as *ArchDaily* and *Dezeen*, the taxation records of not-for-profit research organizations, and the gossip pages of the New York Social Diary. With such a plethora of material we chose not to overwhelm the bibliography and have not included sources where the endnote offers readers their easy retrieval online.

Adkins, Lisa, and Celia Lury. "The Labor of Identity: Performing Identities, Performing Economies." *Economy and Society* 28, no. 4 (1999): 598–614.

Alderman, Derek H. "Place, Naming and the Interpretation of Cultural Landscapes." In *The Ashgate Research Companion to Heritage and Identity*, edited by Brian Graham and Peter Howard, 195–213. Farnham, UK: Ashgate, 2008.

Allen, Thomas, and Gunter Henn. *The Organization and Architecture of Innovation: Managing the Flow of Technology*. Burlington, MA: Elsevier, 2007.

Alvarez, Walter. *T-Rex and the Crater of Doom*. Princeton, NJ: Princeton University Press, 1997.

Baine, Mary R., and Rodney M. Baine. "Blake's Other Tigers, and 'The Tyger.'" *Studies in English Literature 1500–1900* 15, no. 4 (1975): 563–578.

Ball, Philip. *Shapes: Nature's Patterns, A Tapestry in Three Parts*. Oxford: Oxford University Press, 2009.

Barnet, Belinda, and Niles Eldredge. "Material Cultural Evolution: An Interview with Niles Eldredge." *Fibreculture Journal* 3 (2004). http://three.fibreculturejournal.org/fcj-017-material-cultural-evolution-an-interview-with-niles-eldredge/.

Barthes, Roland. *Image-Music-Text*. Translated by Stephen Heath. London: Fontana, 1977.

Barthes, Roland. *The Eiffel Tower and Other Mythologies*. Translated by Richard Howard. Berkeley: University of California Press, 1997.

Basbanes, Nicholas. *A Gentle Madness*. New York: Henry Holt, 1995.

Bataille, Georges. *The Accursed Share, Volume 1*. Translated by Robert Hurley. New York: Zone Books, 1989.

Beesley, Philip, and Sarah Bonnemaison, eds. *On Growth and Form: Organic Architecture and Beyond*. Toronto: Riverside Architectural Press, 2008.

Benjamin, Walter. *One Way Street and Other Writings*. Translated by E. Jephcott and K. Shorter. London: Verso, 1985.

Benjamin, Walter. *Selected Writings (Volume 2 1927–1934)*. Edited by Michael W. Jennings, Howard Eiland, and Gary Smith. Translated by Rodney Livingstone and others. Cambridge, MA: Belknap Press of Harvard University, 1990.

Birke, Lynda, Arnold Arluke, and Mike Michael. *The Sacrifice: How Scientific Experiments Transform Animals and People*. West Lafayette, IN: Purdue University Press, 2007.

Bonner, John Tyler. "Analogies in Biology." *Synthese* 15, no. 2 (1963): 275–279.

Bonner, John Tyler. "Introduction." In D'Arcy Wentworth Thompson, *On Growth and Form*, canto, pp. xiv-xxi. (Cambridge, UK: Cambridge University Press, 2004).

Bonner, John Tyler. *Why Size Matters: From Bacteria to Blue Whales*. Princeton, NJ: Princeton University Press, 2006.

Bonner, John Tyler. *Randomness in Evolution*. Princeton, NJ: Princeton University Press, 2013.

Bove, Paul A. "The End of Humanism: Michel Foucault and the Power of Disciplines." *Humanities in Society* 3 (Winter 1980): 23–40.

Bowler, Peter J. *The Eclipse of Darwinism: Anti-Darwinian Evolution Theories in the Decades around 1900*. Baltimore, MD: John Hopkins University Press, 1983.

Boym, Svetlana. *The Future of Nostalgia*. New York: Basic Books, 2001.

Brown, Andrew. *In the Beginning Was the Worm: Finding the Secrets of Life in a Tiny Hermaphrodite*. London: Simon and Schuster, 2003.

Brown, James Robert. *Who Rules in Science? An Opinionated Guide to the Wars*. Cambridge, MA: Harvard University Press, 2001.

Browne, Janet. *Charles Darwin: The Power of Place: Volume 2 of a Biography*. London: Jonathan Cape, 2003.

Brownell, Blaine. *Hypernatural: Architecture's New Relationship with Nature*. Princeton, NJ: Princeton Architectural Press, 2015.

Brullet, Manel, Albert de Pineda, and Alfonso de Luna. *Parc de Recerca Biomédica de Barcelona*. Barcelona: Edicions de l'Eixample, 2007.

Buffon, Georges-Louis Leclerc, Comte de. *Natural History Abridged*. London: C. and G. Kearsley, 1791.

Buffon, Georges-Louis Leclerc, Comte de. *Histiore naturelle, Quadrupédes*, vol. 3. Paris: A la Librairie Stéréotype de P. Didot, 1799.

Bullock, Alan, and Stephen Trombley, eds. *The Norton Dictionary of Modern Thought*. New York: W. W. Norton, 1999.

Bupta, Vishal, Shailendra Singh, and Naresh Khatri. "Creativity in Research and Development Laboratories: A New Scale for Leader Behaviours." *IIMB Management Review* 25, no. 2 (2013): 83–90.

Burkert, Walter. *Greek Religion*. Translated by John Raffan. Cambridge, MA: Harvard University Press, 1985.

Caillois, Roger. *The Mask of Medusa*. Translated by G. Ordish. New York: Clarkson N. Potter, 1964.

Caillois, Roger. *The Writing of Stones* (1970). Introduction by Marguerite Yourcenar. Translated by Barbara Bray. Charlottesville: University Press of Virginia, 1985.

Canguilhem, Georges. *The Normal and the Pathological*. Translated by Carolyn R. Fawcett and Robert S. Cohen. New York: Zone Books, 1991.

Chamovitz, Daniel. *What a Plant Knows*. New York: Scientific American, 2013.

Clark, Karl, and Stephen Ekker, "How Zebrafish Genetics Informs Human Biology." *Nature Education* 8, no. 4, (2015): 3.

Cohen, Jeffrey Jerome, ed. *Animal, Vegetable, Mineral: Ethics and Objects*. Washington, DC: Punctum Books, 2012.

Colebrook, Claire. *Death of the Post-Human: Essays on Extinction*. Ann Arbor, MI: Open Humanities Press, 2013.

Collins, Peter. *Changing Ideals in Modern Architecture, 1750–1950*. Montreal: McGill-Queens University Press. 1998.

Coote, Jeremy. "Origins of the Afro Comb: 6,000 Years of Culture, Politics, and Identity (review)." *African Arts* 50, no. 4 (2017): 81–82.

Corner, George W. *A Biographical Memoir of Warren Harmon Lewis, 1870–1964*. Washington, DC: National Academy of Sciences, 1967.

Corning, Peter. *The Synergism Hypothesis: A Theory of Progressive Evolution*. New York: McGraw Hill, 1983.

Correa, Charles. *Charles Correa*. London: Thames & Hudson, 1996.

Correa, Charles. "Museums: An Alternate Typology." *Daedalus* 128, no. 3 (1999): 327–332.

Correa, Charles. *A Place in the Shade: The New Landscape and Other Essays*. Berlin: Hatje Cantz, 2012.

Couture, Jean-Pierre. "Sphere-Effects: A History of Peter Slöterdijk's Political Architectures." *World Political Science Review* 7, no. 1 (2011): 1–17. https://doi.org/10.2202/1935-6226.1108.

Crosbie, Michael. "Add and Subtract." *Progressive Architecture* 74, no. 10 (1993): 48–51.

Davenport, Charles B. "The Effects of Race Intermingling." *Proceedings of the American Philosophical Society* 56, no. 4 [1917]: 366–367.

Darwin, Charles. *On the Origin of Species by Means of Natural Selection, or the Preservation of Favoured Races in the Struggle for Life* (1859). Facsimile edition, introduced by Ernst Mayr. Cambridge, MA: Harvard University Press, 1966.

Dawkins, Richard. *The Selfish Gene*. Oxford: Oxford University Press, 1976.

Deleuze, Gilles. *Cinema 1: The Movement-Image*. Translated by Hugh Tomlinson and Barbara Habberjam. Minneapolis, MN: Althone Press, 1986.

Deleuze, Gilles. *The Logic of Sense*. Translated by Mark Lester and Charles Stivale. London: Athlone Press, 1990.

Deleuze, Gilles. "Postscript on the Societies of Control." *October* 59 (Winter 1992): 3–7.

Deleuze, Gilles. *Essays Critical and Clinical*. Translated by Daniel Smith and Michael Greco. Minneapolis: University of Minnesota Press, 1997.

Deleuze, Gilles, and Felix Guattari. *Anti-Oedipus: Capitalism and Schizophrenia* [1972]. Preface by Michel Foucault. Translated by Robert Hurley, Mark Steem, and Helen R. Lane. Minneapolis: University of Minnesota Press, 1983.

Deleuze, Gilles, and Feliz Guattari. *What Is Philosophy?* Translated by Hugh Tomlinson and Graham Burchill. London: Verso, 1994.

Deleuze, Gilles, and Felix Guattari. *A Thousand Plateaus: Capitalism and Schizophrenia* [1980], Translated and Foreword by Brian Massumi. London: Continuum, 2004.

Derrida, Jacques. *Glas* (1974). Translated by John P. Leavey and Richard Rand. Lincoln: University of Nebraska Press, 1986.

Derrida, Jacques, and Maurizio Ferraris. *A Taste for the Secret*. Translated by Giacomo Donis. Cambridge, UK: Polity Press, 2001.

de Saussure, Ferdinand. *Course in General Linguistics*. Translated by Roy Harris. Edited by Charles Bally and Albert Sechehaye. La Salle, IL: Open Court, 1983.

Desmuelles, G., G. Querrec, P. Redou, S. Kordelo, L. Misery, V. Rodin, and J. Tisseau. "The Virtual Reality Applied to the Biology Understanding: The In Virtuo Experimentation." *Expert Systems with Applications* 30, no. 1 (2006): 82–92.

Dewey, John. *Experience and Nature*. Chicago: Open Court, 1925.

Dierig, Sven. "Engines for Experiment: Laboratory Revolution and Industrial Labor in the Nineteenth Century City." *Osiris* 18 (2003): 116–134.

Dostoevsky, Fyodor. *Notes from Underground*. Translated by Constance Garnett. New York: Macmillan, 1918.

Dostoevsky, Fyodor. *Winter Notes on Summer Impressions*. Translated by David Patterson. Evanston, IL: Northwestern University Press, 1997.

Douglas, Angela E. *The Symbiotic Habit*. Princeton, NJ: Princeton University Press, 2010.

Doyle, Richard. *On Beyond Living: Rhetorical Transformations of the Life Sciences*. Stanford, CA: Stanford University Press, 1997.

Doyle, Richard. *Darwin's Pharmacy: Sex, Plants, and the Noosphere*. Seattle: University of Washington Press, 2013.

Duffy, Francis, and Jack Tanis. "A Vision of the New Workplace." *Site Selection and Industrial Development* 162, no. 2 (1993): 427–443.

Durkheim, Emile. *Emile Durkheim: Selected Writings*. Edited by Anthony Giddens. Cambridge, UK: Cambridge University Press, 1972.

Ebert, James D. "Evolving Institutional Patterns for Excellence: A Brief Comparison of the Organization and Management of the Cold Spring Harbor Laboratory and the Marine Biological Laboratory." *Biological Bulletin* 168 (supplement; 1985): 183–186.

Eldredge, Niles, and Stephen Jay Gould. "Punctuated Equilibria: An Alternative to Phyletic Gradualism." In *Models in Paleobiology*, edited by T. J. M. Schopf, 82–115. San Francisco: Freeman Cooper, 1972.

Ellsworth, Elizabeth, and Jamie Kruse. "Introduction." In *Making the Geologic Now: Responses to Material Conditions of Contemporary Life*. Brooklyn, NY: Punctum Books, 2013.

Endersby, Jim. *A Guinea Pig's History of Biology: The Plants and Animals Who Taught Us the Facts of Life*. London: William Heinemann, 2007.

Ferry, Georgina. "The Art of Laboratory Design." *Nature* 457, no. 541 (January 29, 2009): 541. http://www.nature.com/nature/journal/v457/n7229/full/457541a/html.

Feyerabend, Paul. *Against Method: Outline of an Anarchistic Theory of Knowledge*. New York: New Left Books, 1975.

Foucault, Michel. *The Order of Things: An Archaeology of the Human Sciences* [1966]. Translated by Alan Sheridan. New York: Random House, 1970.

Foucault, Michel. *Speech Begins after Death*. Translated by Robert Bononno. Minneapolis: Minnesota University Press, 2013.

Frasca, Robert, and Joseph Giovannini. *Future Tense: Zimmer Gunsul Frasca*. South Pasadena, CA: Balcony Press, 2008.

Fulford, Roger. *Hanover to Windsor*. London: Fontana-Collins, 1966.

Fujimura, Joan. "Technobiological Imaginaries: How Do Systems Biologists Know Nature?" In *Knowing Nature: Conversations at the Intersection of Political Ecology and Science Studies*, edited by M. J. Goldman, P. Nadasdy, and M. D. Turner, 66–80. Durham, NC: Duke University Press, 2011.

Galison, Peter. *Image and Logic: A Material Culture of Microphysics*. Chicago: University of Chicago Press, 1997.

Galison, Peter. "Trading with the Enemy." In *Trading Zones and Interactional Expertise*, edited by Michael E. Gorman. Cambridge, MA: MIT Press, 2010.

Gauer, James. "Medicine Chest: UBC Faculty of Pharmaceutical Sciences." *Architectural Record*, May 16, 2013. http://www.architecturalrecord.com/articles/7936-ubc-faculty-of-pharmaceutical-sciences.

Gehry, Frank, and Nancy E. Joyce. *Building Stata: The Design and Construction of Frank O. Gehry's Stata Center at MIT*. Cambridge, MA: MIT Press, 2004.

Gibbons, Ann. "The Salk Institute at a Crossroads." *Science* 249, no. 4967 (1990): 360–362.

Gieryn, Thomas F. "What Buildings Do." *Theory and Society* 31, no. 1 (2002): 35–74.

Gilbert Scott, George. *Remarks on Secular and Domestic Architecture, Present and Future*. London: John Murray, 1857.

Gissen, David. *Subnature: Architecture's Other Environments*. Princeton, NJ: Princeton Architectural Press, 2009.

Goldberger, Paul. "Imitation That Doesn't Flatter." *New York Times*, April 28, 1996.

Goldhagen, Sarah. "Seeing the Building for the Trees." *New York Times*, January 7, 2012.

Goldstein, Jan. "Foucault among the Sociologists: The 'Disciplines' and the History of the Professions." *History and Theory* 23, no. 2 (1984): 170–192.

Gould, Stephen Jay. "D'Arcy Thompson and the Science of Form." *New Literary History* 2, no. 2, (1971): 229–258.

Gould, Stephen Jay. *The Lying Stones of Marrakech: Penultimate Reflections in Natural History*. New York: Harmony Books, 2000.

Grant, John E. "The Art and the Argument of 'The Tyger.'" In *Discussions of William Blake*, edited by John E. Grant, 76–81. Boston: Heath, 1961.

Gray, Denis, and Umut Toker. "Innovation Spaces: Workspace Planning and Innovation in U.S. University Research Centers." *Research Policy* 37, no. 2 (2008): 309–329.

Gross, Paul, and Norman Levitt. *Higher Superstition: The Academic Left and Its Quarrels with Science*. Baltimore, MD: John Hopkins University Press, 1998.

Gross, Paul, Norman Levitt, and Martin Lewis, eds. *The Flight from Science and Reason*. New York: New York Academy of Sciences, 1996.

Grube, Oswald. "The Birth of the Modern Research Building in the USA." In *Principles of Research and Technology Buildings*, edited by Hardo Braun and Dieter Grömling, 21–26. Basel: Birkhauser, 2005.

Gutman, Robert, and Barbara Westergaard. "Building Evaluation, User Satisfaction, and Design." In *Designing for Human Behaviour: Architecture and the Behavioural Sciences*, edited by Jon T. Lang, Charles Burnette, Walter Moleski and David Vachon, 320–329. Stroudsburg, PA: Dowden, Hutchinson, and Ross, 1974.

Hall, Matthew. *Plants as Persons*. Albany, NY: State University of New York Press, 2010.

Hannaway, Owen. "Laboratory Design and the Aim of Science: Andreas Libavius versus Tycho Brahe." *Isis* 77, no. 4 (1986): 609–610.

Harré, Rom. *Pavlov's Dogs and Schrödinger's Cat: Scenes from the Living Laboratory*. Oxford: Oxford University Press, 2009.

Hawken, Paul, ed. *Drawdown: The Most Comprehensive Plan Ever Proposed to Reverse Global Warming*. New York: Penguin, 2017.

Hensel, Michael. *Emergent Technologies and Design: Towards a Biological Paradigm for Architecture*. Abingdon, UK: Routledge, 2010.

Hillgartner, Stephen, Clark A. Miller, and Rob Hagendijk, eds. *Science and Democracy: Making Knowledge and Making Power in the Biosciences and Beyond*. London: Routledge, 2015.

Hillier, Bill, and Allen Penn. "Visible Colleges: Structure and Randomness in the Place of Discovery." *Science in Context* 4, no.1 (1991): 23–49.

Hodgkin, Jonathan. "Early Worms." *Genetics* 121 (1989): 1–3.

Hollier, Denis. *Against Architecture: The Writings of Georges Bataille*. Translated by Betsy Wing. Cambridge MA: MIT Press, 1989.

Howe, Kerstin et al. "The Zebrafish Reference Genome Sequence and its Relationship to the Human Genome." *Nature* 496, no. 7446 (2013): 498–503.

Hughes, Russell. "The Internet of Politicised 'Things': Urbanisation, Citizenship, and the Hacking of New York 'Innovation' City." *Interstices* 16 (2016): 24–30.

Hyman, Stanley. *The Tangled Bank: Darwin, Marx, Frazer and Freud as Imaginative Writers*. New York: Atheneum, 1962.

Ingraham, Catherine. *Architecture, Animal, Human*. London: Routledge, 2006.

Jameson, Fredric. *The Political Unconscious: Narrative as a Socially Symbolic Act*. Ithaca, NY: Cornell University Press, 1981.

Jardine, Nicholas, and Marina Frasca-Spada. "Splendours and Miseries of the Science Wars." *Studies in History and Philosophy of Science* 28, no. 2 (1997): 219–35.

Jehle-Schulte Strathaus, Ulrike, ed. *Novartis Campus—Fabrikstraße 22: David Chipperfield*. Basel: Christoph Merian Verlag, 2011.

Jencks, Charles. *Modern Movements in Architecture*. London: Pelican, 1973.

Jencks, Charles. "The Iconic Building Is Here to Stay." In *Hunch: The Berlage Institute Report* 11 (Winter 2006/2007): 48–61.

Kaji-O'Grady, Sandra. "The Spaces of Experimental Science." *Journal Spéciale'Z* 4 (2012): 88–99.

Kaji-O'Grady, Sandra, Chris L. Smith, and Russell Hughes. *Laboratory Lifestyles: The Construction of Scientific Fictions*. Cambridge, MA: MIT Press, 2018.

Keane, Sam. *The Violinist's Thumb: And Other Lost Tales of Love, War, and Genius, as Written by Our Genetic Code*. New York: Little, Brown, 2012.

Kisacky, Jeanne. "History and Science: Julien-David Leroy's 'Dualistic Method of Architectural History.'" *Journal of the Society of Architectural Historians* 60, no. 3 (2001): 260–289.

Kitcher, Philip. "A Plea for Science Studies." In *A House Built on Sand: Exposing Postmodernist Myths about Science*, edited by Noretta Koertge, 32–56. Oxford: Oxford University Press, 1998.

Kleinman, Daniel Lee, and Stephen P. Vallas. "Science, Capitalism, and the Rise of the 'Knowledge Worker': The Changing Structure of Knowledge Production in the United States." *Theory and Society* 30, no. 4 (2001): 451–492.

Knorr Cetina, Karin. "The Couch, the Cathedral, and the Laboratory: On the Relationship between Experiment and Laboratory in Science." In *Science as Practice and Culture*, edited by Andrew Pickering, 113–138. Chicago: University of Chicago Press, 1992.

Kohler, Robert. "Labscapes: Naturalizing the Lab." *History of Science* 40, no. 4 (2002): 473–500.

Kohler, Robert. "Lab History: Reflections." *Isis* 99, no. 4 (2008): 761–768.

Kohler, Robert. *Partners in Science: Foundations and Natural Scientists, 1900–1945*. Chicago: University of Chicago Press, 1991.

Kohn, Eduardo. *How Forests Think: Toward an Anthropology beyond the Human*. Berkeley: University of California Press, 2013.

Kuhn, Thomas. *The Structure of Scientific Revolutions*. Chicago: University of Chicago Press, 1962.

Kulper, Amy Catania. "Architecture's Lapidarium: On the Lives of Geological Specimens." In *Architecture in the Anthropocene: Encounters among Design, Deep Time, Science and Philosophy*, edited by Etienne Turpin, 87–110. Ann Arbor, MI: Open Humanities Press, 2013.

Lacan, Jacques. *Seminar III, The Psychoses*. Translated by R. Grigg. London: Routledge/Norton, 1993.

Lampugnani, Vittorio Magnano. "Insight versus Entertainment: Untimely Meditations on the Architecture of Twentieth-Century Art Museums." In *A Companion to Museum Studies*, edited by Sharon MacDonald, 245–262. Malden, MA: Blackwell, 2006.

Latour, Bruno. "Give Me a Laboratory and I Will Raise the World." In *Science Observed: Perspectives on the Social Study of Science*, edited by Karin D. Knorr-Cetina and Michael Mulkay, 141–170. London: Sage, 1983.

Latour, Bruno. "The Costly Ghastly Kitchen." In *The Laboratory Revolution in Medicine*, edited by Andrew Cunningham and Perry Williams, 295–302. Cambridge, UK: Cambridge University Press, 1992.

Latour, Bruno. "To Modernize or to Ecologize? That's the Question." In *Remaking Reality: Nature at the Millennium*, edited by N. Castree and B. Willems-Brawn, 221–241. New York: Routledge, 1997.

Latour, Bruno. "From the World of Science to the World of Research?" *Science* 280, no. 5361, (1998): 208–209.

Latour, Bruno. *Pandora's Hope: Essays on the Reality of Science Studies*. Cambridge, MA: Harvard University Press, 1999.

Latour, Bruno. "What If We Talked Politics a Little?" *Contemporary Political Theory* 2, no. 2 (2003): 143–164.

Latour, Bruno. *Reassembling the Social: An Introduction to Actor Network Theory*. Oxford: Oxford University Press, 2005.

Latour, Bruno. "A Textbook Case Revisited—Knowledge as a Mode of Existence." In *The Handbook of Science and Technology Studies*, 3rd ed., edited by E. Hackett, O. Amsterdamska, M. Lynch, and J. Wacjman, 83–112. Cambridge, MA, MIT Press, 2007.

Latour, Bruno. *On the Modern Cult of the Factish Gods*. Durham, NC: Duke University Press, 2010.

Latour, Bruno, and Paolo Fabbri. "La Rhétorique de la Science Pouvoir et Devoir dans un Article de Science Exacte." *Actes de la Recherche en Sciences Sociales* 13 (1977): 81–95.

Latour, Bruno, and Paolo Fabbri. "The Rhetoric of Science: Authority and Duty in an Article from the Exact Science (1977)." Translated by Sarah Cummins. *Technostyle* 16, no. 1 (2000): 115–134.

Latour, Bruno, and Steve Woolgar. *Laboratory Life: The Construction of Scientific Facts*. Princeton, NJ: Princeton University Press, 1986.

Lazzarato, Maurizio. "Immaterial Labor." In *Radical Thought in Italy: A Potential Politics*, edited by Paolo Virno and Michael Hardt, translated by Paul Colilli and Ed Emery, 133–148. Minneapolis: University of Minnesota Press, 1996.

Lazzarato, Maurizio. "From Capital-Labour to Capital-Life." Translated by Valerie Fournier, Akseli Vertanen and Jussi Vähämäki. *Ephemera: Theory and Politics in Organization* 4, no. 3 (2004): 177–208.

Lazzarato, Maurizio. "'Exiting Language,' Semiotic Systems and the Production of Subjectivity in Félix Guattari." In *Cognitive Architecture: From Bio-politics to Noo-politics*, translated by Eric Anglés, edited by Deborah Hauptmann and Warren Neidich, 502–521. Rotterdam: 010 Publishers, 2010.

Lazzarato, Maurizio. *Signs and Machines: Capitalism and the Production of Subjectivity*. Translated by Joshua Jordan. Los Angeles: Semiotext(e), 2014.

Leslie, Stuart. "'A Different Kind of Beauty': Scientific and Architectural Style in I. M. Pei's Mesa Laboratory and Louis Kahn's Salk Institute." *Historical Studies in the Natural Sciences* 38, no. 2, (2008): 173–221.

Lessard, Suzannah. *The Architect of Desire: Beauty and Danger in the Stanford White Family*. New York: Delta, 1991.

Lévi-Strauss, Claude. *Totemism*. Translated by Robert Needham. Boston: Beacon Press, 1963.

Lévi-Strauss, Claude. *Introduction to the Work of Marcel Mauss*. Translated by Felicity Baker. London: Routledge and Kegan Paul, 1987.

Lim Ee Man, Joseph. *Bio-Structural Analogues in Architecture*. Amsterdam: BIS Publishers, 2009.

Livingstone, David. *Putting Science in Its Place: Geographies of Scientific Knowledge*. Chicago: University of Chicago Press, 2003.

Liu, Alan. *The Laws of Cool: Knowledge Work and the Culture of Information*. Chicago: University of Chicago Press, 2004.

Lynn, Greg. *Animate Form*. Princeton, NJ: Princeton University Press, 1998.

Lyotard, Jean-Francois. *The Postmodern Condition: A Report on Knowledge*. Translated by Geoff Bennington and Brian Massumi. Manchester, UK: Manchester University Press, 1984.

Marder, Michael. *Plant-Thinking: A Philosophy of Vegetal Life*. New York: Columbia University Press, 2013.

Martin, Reinhold. *The Organizational Complex: Architecture, Media and Corporate Space*. Cambridge, MA: MIT Press, 2003.

Massey, Jonathan. "Ornament and Decoration." In *The Handbook of Interior Architecture and Design*, Part 3.1 *Atmospheric Conditions of the Interior*. Edited by Graeme Brooker and Lois Weinthal, 497–513. London: Bloomsbury, 2013.

Mauss, Marcel. *A General Theory of Magic*. Translated by Robert Brain. London: Routledge, 2001.

Mayer, Jed. "Ways of Reading Animals in Victorian Literature, Culture and Science." *Literature Compass* 7, no. 5 (2010): 347–357.

McElheny, Victor K. *Watson and DNA: Making a Scientific Revolution*. New York: Perseus/Wiley, 2003.

McGurn Centellas, Katherine. "For Love of Land and Laboratory: Nation-Building and Bioscience in Bolivia." PhD diss., University of Chicago, 2008.

McKinlay, Alan, and Philip Taylor. "Power, Surveillance and Resistance: Inside the 'Factory of the Future.'" In *The New Workplace and Trade Unionism*, edited by P. Ackers, C. Smith, and P. Smith, 279–300. London: Routledge, 1996.

McRobbie, Angela. *Be Creative: Making a Living in the New Culture Industries*. London: Polity, 2016.

Mees, C. E. Kenneth. "The Production of Scientific Knowledge." *Science* 46, no. 1196 (November 30, 1917): 519–528.

Miller, Elaine. *The Vegetative Soul*. Albany: State University of New York Press, 2002.

Mitchell, Robert. *Experimental Life: Vitalism in Romantic Science and Literature*. Baltimore, MD: Johns Hopkins University Press, 2013.

Mooney Nickel, Patricia. "Haute Philanthropy: Luxury, Benevolence and Value." *Luxury* 2, no. 2 (2016): 11–31.

Morton, Timothy. *Ecology without Nature: Rethinking Environmental Aesthetics*. Cambridge, MA: Harvard University Press, 2009.

Negroponte, Nicholas. *Architecture Machine*. Cambridge, MA: MIT Press, 1970.

Norberg-Schulz, Christian. *Genus Loci: Toward a Phenomenology of Architecture*. New York: Rizzoli, 1991.

Owen, Richard. *The Archetype and Homologies of the Vertebrate Skeleton*. London: John Van Voorst, 1848.

Pallasmaa, Juhani. *The Eyes of the Skin: Architecture and the Senses*. Preface by Steven Holl. Chichester, UK: Wiley Academy, 2005.

Paul, Steven M., Daniel Mytelka, Christopher Dunwiddie, Charles Persinger, Bernard Munos, Stacy Linborg and Aaron Schacht. "How to Improve R&D Productivity: The Pharmaceutical Industry's Grand Challenge." *Nature Reviews: Drug Discovery* 9 (March 2010): 203–214.

Pawley, Emily. "'The Balance Sheet of Nature': Calculating the New York Farm, 1820–1860." PhD diss., University of Pennsylvania, 2009.

Picon, Antoine. *Ornament: The Politics of Architecture and Subjectivity*. London: John Wiley and Sons, 2013.

Pollen, Michael. *The Botany of Desire: A Plant's-Eye View of the World*. New York: Random House, 2001.

Porter, Tom. *Will Alsop—The Noise*. London: Routledge, 2011.

Prusinkiewicz, Przemyslaw. "Art and Science of Life: Designing and Growing Virtual Plants with L-Systems." In *Nursery Crops: Development, Evaluation, Production and Use: Proceedings of the XXVI International Horticultural Congress*, edited by C. Davidson and T. Fernandez, *Acta Horticulturae (ISHS)* 630, (2004): 15–28.

Rabinow, Paul, and Talia Dan-Cohen. *A Machine to Make a Future: Biotech Chronicles*. Princeton, NJ: Princeton University Press, 2005.

Raine, Kathleen. "Who Made the Tyger?" *Encounter 2, no. 6,* (June 1954): 43–49.

Roberts, Leslie. "The Worm Project." *Science* 249, no. 4961 (June 15, 1990): 1310–1313.

Rogoff, Irit. "Gossip as Testimony: A Postmodern Signature." In *Generations and Geographies in the Visual Arts: Feminist Readings*, edited by Griselda Pollock, 58–65. New York: Routledge, 1996.

Rousseau, Henri. *The Collected Writings of Rousseau*. Edited by Roger D. Masters and Christopher Kelly. Hanover, NH: University Press of New England, 1990.

Rubin, Gerald. "Janelia Farm: An Experiment in Scientific Culture." *Cell* 125, no. 2 (2006): 209–212.

Rudwick, Martin J. S. *Scenes from Deep Time: Early Pictorial Representations of the Prehistoric World*. Chicago: University of Chicago Press, 1992.

Rudwick, Martin J. S. *Georges Cuvier, Fossil Bones, and Geological Catastrophes*. Chicago: University of Chicago Press, 1997.

Rusch, D. B., A. L. Halpern, G. Sutton et al. "The Sorcerer II Global Ocean Sampling Expedition: Northwest Atlantic through Eastern Tropical Pacific." *PLoS Biology* 5, no. 3 (2007): e77, http://journals.plos.org/plosbiology/article?id=10.1371/journal.pbio.0050077.

Ruskin, John. *The Stones of Venice*, volumes 1–3 [1851–1853]. Edited by J. G. Links. New York: Da Capo Press, 2003.

Scheerbart, Paul, and Bruno Taut. *Glass Architecture*. New York: Praeger, 1972.

Schmertz, Mildred, and Deborah Dietsch. *Zimmer Gunsul Frasca: Building Community*. Rockport, MA: Rockport Publishers, 1995.

Schmidgen, Henning. *Bruno Latour in Pieces*. Translated by Gloria Custance. New York: Fordham University Press, 2014.

Scott, Geoffrey. *The Architecture of Humanism: A Study in the History of Taste*. London: Methuen, 1914.

Searle, John. *The Construction of Social Reality*. New York: Free Press, 1995.

Semper, Gottfried. "Development of Architectural Style." *Inland Architect and News Record*, December 1889, 76–78; January 1890, 92–94; February 1890. 5–6; March 1890, 32–33.

Semper, Gottfried. *Style in the Technical and Tectonic Arts, Or, Practical Aesthetics*. Introduction by Harry Francis Mallgrave. Translation by Harry Francis Mallgrave and Michael Robinson. Los Angeles: Getty Research Institute, 2004.

Semper, Gottfried. *Du Style et de l'architecture: Écrits, 1834–1869*. Marseille: Parenthèses, 2007.

Semple, Janet. *Bentham's Prison: A Study of the Panoptican Penitentiary*. Oxford: Clarendon, 1993.

Serres, Michel. *Hermes: Literature, Science, Philosophy*. Baltimore, MD: Johns Hopkins University Press, 1982.

Serres, Michel. "Literature and the Exact Sciences." *Substance: A Review of Theory and Literary Criticism* 18, no. 2 (1989): 3–34.

Serres, Michel, and Bruno Latour. *Conversations on Science, Culture, and Time.* Translated by Roxanne Lapidus. Ann Arbor: University of Michigan Press, 1995.

Shapin, Steven. "'The Mind Is Its Own Place': Science and Solitude in Seventeenth-Century England." *Science in Context* 4, no. 1 (1991): 191–218.

Shapin, Steven. *A Social History of Truth: Civility and Science in Seventeenth-Century England.* Chicago: University of Chicago Press, 1994.

Shapin, Steven. "I'm a Surfer." *London Review of Books* 30, no. 6 (2008): 5–8. http://www.lrb.co.uk/v30/n06/steven-shapin/im-a-surfer.

Shapin, Steven. *Never Pure: Historical Studies of Science as If It Was Produced by People with Bodies, Situated in Time, Space, Culture, and Society, and Struggling for Credibility and Authority.* Baltimore, MD: John Hopkins University Press, 2010.

Shapin, Steven, and Simon Schaffer. *Leviathan and the Air Pump: Hobbes, Boyle and the Experimental Life.* Princeton, NJ: Princeton University Press, 1985.

Silverman, Kaja. *The Subject of Semiotics.* New York: Oxford University Press, 1983.

Simondon, Gilbert. *On the Mode of Existence of Technical Objects* [1958]. Translated by Cécile Malaspina and John Rogove. Minneapolis: Univocal Publishing, 2017.

Sloterdijk, Peter. "The Crystal Palace" [2005]. Translated by Michael Darroch. *Public* 37 (2008): 11–16. https://pi.library.yorku.ca/ojs/index.php/public/article/viewFile/30252/27786.

Sloterdijk, Peter. *Spheres II–Globes, Macrospherology.* Translated by Weiland Hoban. South Pasadena, CA: Semiotext(e), 2014.

Sloterdijk, Peter. *Spheres III–Foams, Plural Spherology.* Translated by Weiland Hoban. South Pasadena: Semiotext(e), 2016.

Smith, Chris L., and Sandra Kaji-O'Grady. "Exaptive Translations between Biology and Architecture." *Architecture Research Quarterly* 18, no. 2 (2014): 155–166.

Smith, Chris L. *Bare Architecture: A Schizoanalysis*, London: Bloomsbury, 2017.

Sokal, Alan, and Jean Bricmont. *Fashionable Nonsense: Postmodern Abuse of Science.* New York: Picador, 1999.

Sokal, Alan, M. Lynch, and S. Cole. "Science and Technology Studies on Trial: Dilemmas of Expertise." *Social Studies of Science* 35, no. 3 (2005): 269–311.

Solomon, Lewis D. *Synthetic Biology: Science, Business, and Policy.* Piscataway, NJ: Transaction Publishers, 2011.

Soulillou, Jacques. *Le décoratif.* Paris: Klincksieck, 1990.

Soulillou, Jacques. *Le Livre de l'ornement et de la guerre.* Marseille: Parenthèses, 2003.

Speidel, Michelle. "The Parasitic Host: Symbiosis Contra Neo-Darwinism." *Pli* 9, 2000: 119–138.

Stamper, John. "London's Crystal Palace and Its Decorative Iron Construction." In *Function and Fantasy: Iron Architecture in the Long Nineteenth Century*, edited by Paul Dobraszczyk and Peter Sealy, 25–48. London: Routledge, 2016.

Steadman, Philip. *The Evolution of Designs: Biological Analogy in Architecture and the Applied Arts*. Abingdon: Routledge, 2008.

Stein, Karen. "The Salk Institute Remains a Modernist Beacon." *Architectural Digest*, January 31, 2013.

Stengers, Isabelle. *Power and Invention: Situating Science*. Foreword Bruno Latour. Translated by Paul Bains. Minneapolis: University of Minnesota Press, 1997.

Stroll, Avrum. *Surfaces*. Minneapolis: University of Minnesota Press, 1988.

Sulston, John, and Ferry, Georgina. *The Common Thread: A Story of Science, Politics, Ethics, and the Human Genome*. London: Corgi, 2003.

Takashi, K., K. Yugi, K. Hashimoto, Y. Yamada, C. J. F. Pickett, and M. Tomita. "Computational Challenges in Cell Simulation: A Software Engineering Approach." *Intelligent Systems, IEEE* 17, no. 5 (2002): 64–71.

Tarquinio, J. Alex. "Long Island Laboratory's Expansion Hides in (and under) Six Buildings." *New York Times*, June 23, 2009. http://www.nytimes.com/2009/06/24/realestate/commercial/24lab.html.

Thrift, Nigel. "Re-inventing Invention: New Tendencies in Capitalist Commodification." *Economy and Society* 35, no. 2 (2006): 279–306.

Tomita, Masura. "Whole-cell Simulation: A Grand Challenge of the 21st Century." *Trends in Biotechnology* 19, no. 6 (2001): 205–210.

Urbanik, Julie. "Locating the Transgenic Landscape: Animal Biotechnology and Politics of Place in Massachusetts." *Geoforum* 38, (2007): 1205–1218.

Valen, Dustin. "On the Horticultural Origins of Victorian Glasshouse Culture." *Journal of the Society of Architectural Historians* 75, no. 4 (2016): 403–423.

Venter, Craig. *A Life Decoded: My Genome, My Life*. New York: Penguin, 2007.

Venturi, Robert, and Denise Scott Brown. "A Protest concerning the Extension of the Salk Center [1993]." In *Iconography and Electronics Upon a Generic Architecture: A View from the Drafting Room*, Robert Venturi, 81–83. Cambridge MA: MIT Press, 1996.

Venturi, Robert, Steven Izenour, and Denise Scott Brown. *Learning from Las Vegas: The Forgotten Symbolism of Architectural Form*. Revised Edition. Cambridge MA: MIT Press, 1977.

Vidler, Anthony. *The Architectural Uncanny: Essays in the Modern Unhomely*. Cambridge MA and London: MIT Press, 1992.

Waldrop, Mitchell. "Research at Janelia: Life on the Farm." *Nature* 479, no. 7373 (November 16, 2011): 284–286.

Walker, Gabrielle. "The Collector." *New Scientist* 179, no. 2405, (26 July, 2003): 38–41.

Warner, Marina. "The Writing of Stones." *Cabinet* 29, 2008. http://cabinetmagazine.org/issues/29/warner.php.

Watch, Daniel, Deepa Tolat, and Gary McNay. "Academic Laboratory." In *Whole Building Design Guide* (Washington DC: National Institute of Building Sciences, 2010), p. 4 of 7, accessed July 2018. http://www.wbdg.org/design/academic_lab.php.

Watson, Elizabeth L. *Houses for Science: A Pictorial History of Cold Spring Harbor Laboratory*. New York: Cold Spring Harbor Laboratory Press, 1991.

Watson, Elizabeth L. *Grounds for Knowledge: A Guide to Cold Spring Harbor Laboratory's Landscapes and Buildings*. Cold Spring Harbor, NY: Cold Spring Harbor Laboratory Press, 2008.

Watson, James. *The Double Helix: A Personal Account of the Discovery of the Structure of DNA* (1968). New York: W. W. Norton, 1980.

Watson, James. *Genes, Girls, and Gamow: After the Double Helix*. Cold Spring Harbor Laboratory Press/Alfred A. Knopf: New York, 2001.

Watson, James. *Avoid Boring People: Lessons from a Life in Science*. Oxford: Oxford University Press, 2007.

Weder, Adele. "Strong Medicine." *German-Architects Review*, September 3, 2013. http://www.german-architects.com/en/projects/42660_Pharmaceutical_Building_on_the_University_of_British_Columbia_Campus.

Weinstock, Michael. *The Architecture of Emergence: The Evolution of Form in Nature and Civilisation*. London: Wiley, 2010.

Wenner, Melinda. "The Most Transparent Research." *Nature Medicine* 15, (2009): 1106–1109.

White, Richard Mark, Anna Sessa, Christopher Burke, Teresa Bowman, Jocelyn LeBlanc, Craig Ceol, Caitlin Bourque, Michael Dovey, Wolfram Goessling, Caroline Erter Burns, and Leonard I. Zon. "Transparent Adult Zebrafish as a Tool for In Vivo Transplantation Analysis." *Cell Stem Cell* 2, no. 2 (2008): 183–189.

Whiteley, Nigel. "Intensity of Scrutiny and a Good Eyeful: Architecture and Transparency." *Journal of Architectural Education* 56, no. 4 (2003): 8–16.

Whitley, Richard. *The Intellectual and Social Organization of the Sciences*. Oxford: Clarendon Press, 1984.

Whyte, William. *The Organization Man* [1956]. Philadelphia: University of Pennsylvania Press, 2002.

Winsberg, Eric. "Simulated Experiments: Methodology for a Virtual World." *Philosophy of Science* 70, no. 1 (2003): 105–125.

Witkowski, Jan A. and John R. Inglis. *Davenport's Dream: 21st Century Reflections on Heredity and Eugenics*. New York: Cold Spring Harbor Laboratory Press, 2008.

Wulf, William. "The Collaboratory Opportunity." *Science* 261 (August 1993): 854–855.

Zamyatin, Yevgeny. *We*. Translated and with a foreword by Gregory Zilboorg. New York: E. P. Dutton, 1959.

Zamyatin, Yevgeny. *We*. Translated and with an introduction by Clarence Brown. London: Penguin, 1993.

Zare, Richard. "Knowledge and Distributed Intelligence." *Science* 275, no. 5303 (1997): 1047.

ZGF Architects LLP. "J. Craig Venter Institute" (online portfolio). Accessed August 11, 2015. http://issuu.com/zgfarchitectsllp/docs/j._craig_venter_institute?e=5145747/7966125.

Žižek, Slavoj. "Nobody Has to Be Vile." *London Review of Books* 28, no. 7 (2006): 10.

Žižek, Slavoj. "We Don't Want the Charity of Rich Capitalists." *ABC Religion and Ethics*, August 14, 2012. http://www.abc.net.au/religion/articles/2012/08/14/3567719.htm.

Index